STEEL

Design, histories, futures

This series aims to advance knowledge on the wider historical significance of design, and, in doing so, go beyond the current scope of "design history." It will also strive to demonstrate that a historical engagement with design necessitates engagement with the wider crisis of the discipline of history itself.

The contributing authors to the series will no doubt bring very different perspectives to the realization of the series' aim and the intellectual challenges it presents. However, they will all share an understanding of the significance of design thought and design action for sustaining the future well-being of humanity and the environments of our dependence. They will also recognize that for this potentiality to be realized, the scope of historical inquiry has to be significantly widened, become more critical, and surpass the limitations of existing concerns with disciplinary boundaries.

The actual directional consequences of designing and of the designed worlds of human occupation, historically and futurally, are still not adequately understood either in or beyond design education, practice, history and theory. Without understanding design as both historically situated and futurally directional, the ethical character of design—as a negotiation between creation and destruction, care and uncaring—cannot be adequately grasped. This series aims to expand the scope of discourse and comprehension of the directional agency of design, extending understanding and prompting speculation toward this end.

Tony Fry, Lisa Norton and Anne-Marie Willis

Series Editors

TITLES IN THE SERIES

Design and the Question of History by Tony Fry, Clive Dilnot and
Susan C. Stewart
Steel by Tony Fry and Anne-Marie Willis
The Future by Design by Damian White

STEEL: A DESIGN, CULTURAL AND ECOLOGICAL HISTORY

Tony Fry and Anne-Marie Willis

Bloomsbury Academic
An imprint of Bloomsbury Publishing Plc

B L O O M S B U R Y
LONDON • NEW DELHI • NEW YORK • SYDNEY

Bloomsbury Academic

An imprint of Bloomsbury Publishing Plc

50 Bedford Square	1385 Broadway
London	New York
WC1B 3DP	NY 10018
UK	USA

www.bloomsbury.com

BLOOMSBURY and the Diana logo are trademarks of Bloomsbury Publishing Plc

First published 2015

© Tony Fry and Anne-Marie Willis, 2015

Tony Fry and Anne-Marie Willis have asserted their right under the Copyright, Designs and Patents Act, 1988, to be identified as Authors of this work.

British Library Cataloguing-in-Publication Data

A catalogue record for this book is available from the British Library.

ISBN: HB: 978-0-8578-5479-7

PB: 978-0-85785480-3

ePDF: 978-1-4725-9170-8

ePub: 978-0-8578-5481-0

Library of Congress Cataloging-in-Publication Data

Fry, Tony.

Steel : a design, cultural, and ecological history / Tony Fry & Anne-Marie Willis.

pages cm.– (Design, histories, futures)

Includes bibliographical references and index.

ISBN 978-0-85785-479-7 (hardback : alk. paper)– ISBN 978-0-85785-480-3 (pbk. : alk. paper)– ISBN 978-1-4725-9170-8 (ePDF)– ISBN 978-0-85785-481-0 (ePub) 1. Steel–History. I. Willis, Anne-Marie. II. Title.

TA472.F79 2014

669′.142–dc23

2014017385

Series: Design, Histories, Futures

Typeset by Fakenham Prepress Solutions, Fakenham, Norfolk NR21 8NN

Printed and bound in Great Britain

CONTENTS

LIST OF CAPTIONS

FIGURE I.0 Eiffel Tower, icon of nineteenth-century steel engineering. Photo by Tony Fry

INTRODUCTION

We are constantly told that economies are dematerializing and that experience is becoming ever more virtualized. In this context, to embark upon a project which seeks to understand a material, and a commonplace one at that, in all its cultural, technical and historical complexity, might seem like a very unfashionable thing to do. Yet it is this very disappearance from public discourse of that which remains stubbornly, materially present that compels us to insist upon the necessity of a project which forces attention on the increasingly overlooked material substrate of our everyday lives. Our project also contests the material/immaterial dichotomy. In the last few decades, material production has certainly lost its leading edge status, being symbolically dethroned by the rise of the immaterial economy in which image and information are driving the creation of economic value. But while the power of logo/brand name increases, that to which it is attached has not disappeared, but simply slipped out of view as material production has become increasingly decentered and mobile, migrating from "industrialized" to "newly industrializing" regions. While marketers and image-makers in advanced economies work to carefully craft branded identities for sports shoes or personalities for plush toys, the actual stuff that comes to bear the created meanings gets manufactured wherever in the world the cheapest labor can be found. The shift then has not been from material to immaterial production, but rather, that the immaterial, as information, meaning or sign, has come, directly or indirectly, to drive material production. This is in fact not a new development. As we will see, it is just that it has become more obvious in recent times.[1] Understanding the nature of the material/immaterial relation in the current moment is vital for thinking the future of steel or of any other material; grasping the inadequacies of how this relation and how this historical moment is dominantly characterized is also vital.

The methodology of this project is based upon a "relational" approach, which has informed other work we have done on materials and the designed environment.[2] This was developed to account for the impacts of materials in the complexity of their contexts and in response to the rise of more narrow, quantification-based approaches to industrial environmental impact assessment, such as embodied energy analysis.[3] Taking a relational approach to the exploration of a material means not viewing it as discrete or singular. Steel, for example, cannot be considered independently from iron or from carbon-based fuel (charcoal, coal or coke). Nor can it even be assumed that steel is a clearly definable form of matter (while there are well over 20,000 formulations of steel in the marketplace, this figure itself means little because of the capacity for customization). A relational approach is consistent with, but also significantly extends, the way in which environmental impacts are currently understood by the advanced sectors of the steel industry, which is via "life-cycle-analysis," a method which *conceptually* (but rarely in practice) provides the possibility of making connections between environmental impacts across time, geographical space, multiple processes and materials.

Like plastic, the word steel is a common one and most people when they hear or read it conjure an image of a material in their mind. But like plastic, "steel" generically names a wide range of materials as well as having acquired a metaphysical status. Steel is thus taken to be a strong material and a metaphor of strength. This further suggest why a strictly empirical approach to understanding its impacts is not adequate, as does the problem of defining iron and steel historically and cross-culturally.

Steel has been differentially defined over time and there are difficulties in translating words from different languages that refer to different kinds of iron. Metallurgical knowledge is often employed trans-historically and trans-culturally to decide whether or not a particular material is steel—this on the basis of its carbon content, or the differences in carbon content between the material's surface and core; or sometimes it is appearance and performative qualities that are used as criteria to judge. We will not attempt to retrospectively apply these contemporary metallurgical definitions of steel. Instead, we acknowledge that since ancient times, distinctions have been drawn on the basis of qualities such as malleability, hardness, softness, ductility, tensile strength and color, generating complex classificatory systems within which the highest grade of metal that combined the most desired qualities often got designated as "steel."

What are ecologies of steel?

One way we will signal relationality in this text is by talking about particular "ecologies of steel." But there is more than one kind of ecology. Every spatial environment is accompanied by an ecology (a system). Environments are connected and transformed by ecologies (systems relationally connect). Relationality is in fact a very useful way to think about and beyond systems, their internal functions and interactions.

It is no longer appropriate to view ecologies just as natural systems, if it ever was. This is because of the depth and extent of transformations of "the natural" by "the artificial" that have occurred over many thousands of years, but which have gathered pace over the last century to the extent of seeming to erase the line between the two. Genetic engineering is a more recent instance of the breakdown of the natural/artificial binary. However, ecologically speaking, iron pre-dated this binary breakdown by many eons.

Iron is usually deemed inanimate and artificial. Yet iron is the core of our planet; it makes up some 4 percent of its crust, and is also part of the very lifeblood of all red-blooded animals. Iron is a bridge between the inert and the organic; it is an active element that links natural and unnatural ecologies. It is vital for the health of the human body, enabling the manufacture of hemoglobin, which is essential for transporting oxygen to the lungs, brain and all other parts of the body. Iron, via the food chain, is extracted from the land as a mineral trace element absorbed by plants and, thereafter, animals. We ingest iron from meat, vegetables and nuts. Iron then is part of a vast and complex web of ecological relations still only partly comprehended by the natural sciences. Yet this complexity is still only part of the picture, for our ecology of dependence is constituted as much from what we have made as from all those ongoing processes whose origins pre-dated, and have been altered by, human presence. Iron and steel have played, and still play, a major part in ecological formation and transformation.

Iron and steel will be shown to have significantly changed the "nature" of the planet by: the appropriation of material resources; the impact of manufacturing processes; the use of iron and steel in other environmentally transformative activities, from agriculture to arms and transport; the kinds of environments iron and steel have enabled to be constructed, from skyscrapers to underground railways; and by their use in the manufacture of myriad products in almost every space of human existence. The making of the modern world is inseparable from the expansionary production of environments of iron, steel and reinforced concrete, the production of all those objects in iron and steel that became implicated in countless economic and social functions of everyday life and of the body that is materially and immaterially formed in this world. Then of course, there are ecologies of meaning, and here we encounter iron and steel as language, image, symbol, metaphor.

The politics of the project

The stories of "ecologies of steel" can tell something of the impact and future of a particular material of human artifice and something of all materials of our world-making. More than this, in telling the stories of steel we can come to realize that that which we now call unsustainability has been a *telos*, a direction put in place and forcefully driven from the very moment that anthropoidal beings started to shape a world to the needs of dwelling beyond animality. In making environments, in using tools, human beings made both themselves and their fate as technological world transformers. The bringing of fire to metal not only created a quantum leap in the potential for technological advancement, it also accelerated the passage toward the unsustainable. This is not to condemn distant, past generations, on the assumption that they could foresee where their actions would lead. However, once it is realized that the forms of human world-making have been increasingly towards world negation (unsustainability), a moment of ethical confrontation arrives.[4] We humans of this epoch are of this moment. We are the generation that have to change the direction away from unsustainable "growth and development" in order to secure the "being-of-being" (the relational web of being in which we are implicated). The task, in these circumstances, is not "saving the planet" but rather, taking responsibility for what we do individually and collectively. As soon as we see this as an engagement with what we build, how we make, what we learn, and how we dwell, we are back in a world of material fabric of which steel is a big part.

What all this means in direct terms is gaining a much better understanding of the impacts of iron and steel-making in the past and present, as well as how to significantly reduce its impacts in the future. This involves examining what is made with steel, why, and how it could be used more sustainably. But caution is needed here, as "sustainable" has become a very loose term, attached willy-nilly to all kinds of activities which in fact sustain the unsustainable by seeking, or maybe only appearing, to slightly modify the impacts of something which is, in essence, damaging.

We seek to give substance to these claims via this project. We shall be re-reading the history of iron and steel-making in Europe and Asia as a material and cultural archaeology that transformed landscapes, climates, ecologies, industries, infrastructures and ways of life. We will look at the language and culture of iron and steel-making as it played a part in the rise of scientific knowledge—specifically, we will revisit thinkers from antiquity, East and West to examine the *continuity* of metallurgy across what most historians usually miscast as a progression from magic to alchemy to science.

Improving the performance of metal has been an unceasing preoccupation of metallurgy. The history of iron-making is inscribed in the material's present and future. Advances in the development and refinement of iron and steel always trade on knowledge and technologies from the past, but much more is carried from "the past" than the dominant narratives of iron and steel generally acknowledge.

The material research

This project is not just about a specific material, or even about "the material" per se; rather it is a neo-materialist exploration of the determinate relations of steel from the perspective of the relationality of ecologies (as opposed to the essentialism of "the ecological").

This is not the same as "material determination." Simplistic notions of material causality become redundant, as soon as we admit the numerous determinate relations of steel (what determines it and what it determines). This panoply of relations is manifested as exchange between knowledge, materials, technologies, cultures, economies. This is one way of characterizing "the ecologies of steel" and clearly, it opens up a complexity beyond a single system or structure, but which we will strive to grasp and make available to view, at least in part.

Language and perspective

Besides the technical languages of steel and metallurgy, this account draws on histories of technology, science, intercultural studies, environmental studies, as well as design history and theory. This brings advantages and problems. Viewing the object of study from multiple perspectives enables a rich and complex picture to be assembled. But for readers, this means encountering terms that shift between familiar and unfamiliar. The text needs to be read at variable speeds: the new will need to be taken slowly, while the familiar can be moved across quickly. However, caution is needed, because the way in which standard accounts will be treated will not always be standard. The treatment of the history of iron-making in Europe and Asia is a case in point: a revised assessment of how knowledge travels and a different perspective on "development" will create significant differences of historical interpretation.

History is given prominence in our account because we believe that one of the major reasons the condition of unsustainability goes largely unrecognized today is a preoccupation with the present, a looking to "the future" and a forgetting of the past. There are no quick fixes—things cannot become sustainable instantly. Sustainability cannot be created unless the condition of unsustainability is thoroughly understood, and this cannot be done without historical knowledge. So history, as revised, has a very important future.

There is a vast literature on the history of European iron and steel-making. Many of these histories treat the development of iron and steel-making as a series of technological progressions: from early methods of smelting ore in bloomeries to the arrival of the blast furnace and the foundry industry; then the development of steel-making processes, the introduction of the Bessemer converter and open-hearth steel-making, followed by an account of the modern integrated steel works, basic oxygen steel-making and ending with the mini-mill and the electric arc furnace. We engage this history, but our intention is to extend and recast it. For readers wishing to explore standard accounts there are a number referenced in our bibliography.

While such histories of iron and steel give the impression of a single narrative of progress, there really is not just one story or one position of speech. The more those differences between languages, cultures and values have come to be recognized, the more difficult and inappropriate it has become to secure a single account of any historical phenomenon. This does not assume that all perspectives are equal (pluralism) but rather that an ethical choice has to be made in the face of the differences of the plural. In the case of the "history" of iron and steel, a globally integrated account that "pulls together" all available histories into one history would not only fly in the face of this thinking, but be an impossible task—nobody, no thing or event ever arrives cut from context and totalized within a single frame of reference. Because histories are contestable and there is no neutral space from which to tell, all one can do is to make one's viewpoint, one's bias, explicit. Without question, we write with a bias toward sustainment.

These comments connect with how we will be viewing environmental impacts.

Clearly human lives, all life forms, have environmental impacts. We cannot eliminate impacts—that is not the aim. Rather what can be done is to develop a better understanding of consequences and of the difference between positive and negative impacts (briefly, those that sustain ecologies versus those that destroy ecologies, be they biophysical, social or symbolic). This knowledge can then be used to exercise responsibility and make decisions.

Again, environments have to be seen relationally. Our (Western) understanding of what constitutes an environment is part of the problem. We assume that a building site, city, park, forest or garden is discrete, something bounded, whereas ecological relations mostly operate within and across such boundaries in ways quite at odds with our image of them. Our mode of seeing is an historic construct and our knowledge of "the world" is culturally specific. This point has been made many times before, especially in relation to values, behaviors and the domain of the social. Bringing this perspective to the notion of environmental impacts has another implication, which is that we are constantly in a situation of acting, and thus enacting transformations, but in a condition of very limited knowledge. It is not as if the evidence is simply there but hidden, rather it is that we mostly lack the sensibility or disposition to see available signs, think what is not normally thought or speak what is normally silent. Without question, one of the major aims of this book is to help create this sensibility. For this to happen not only does the way we think environments have to change, but also the way we think many other things such as: science, alchemy and magic; cultures as Eastern or Western; the pre-modern, modern and postmodern.

Problems and solutions

There has been a long-standing and, as yet, historically unregistered tension between the creation of the unsustainable and the desire and need for sustainability. Current forms and forces of unsustainability are lodged in long-standing practices, values and thinking. The archaeology of unsustainability is to be found first in the coming to dominance of cultures that viewed the resources of planet Earth as an infinite "standing reserve" to simply use at will. The second historically long-standing factor is the "sustainment of the immediate." In other words, for many cultures, short-term action to sustain the status quo has failed to take into account the need for structural sustainability of all that is essential to sustain (which is itself historically and geographically variable as environments and ecologies change). It has only been in recent times that the problem of unsustainability and the need for the sustainable has arrived. Even so, the nature of both unsustainability and sustainability is still barely understood. Our anthropocentrism (human-centeredness) foregrounds sustainability as the sustainment of the humanoid species, and the human in a web on non-human ecologies. In other words, the making of the crisis of unsustainability is a projection of human needs and values upon material circumstances—it is objective only from our point of (subjective) view. We may eliminate ourselves and many other life forms, but it is extremely unlikely that that we have the ability to obliterate all life. In this context "sustainability" is a value that is attempted to be realized as a material condition to mobilize against the long reach of a propensity towards the unsustainable, which so far, in our limited way, we have only objectified as discernible environmental impacts.

While what has just been outlined is very abstract, one of the main imperatives of this project is to historically concretize these claims by using steel as a case study for considering how the unsustainable might be turned toward the sustainable. The choice of steel is, of course, not arbitrary. In its inseparable relation to iron, it travels back in time and across all continents; as the primary material of industrial production, it has been at the core of the making of the industrialized world—its tools, economies, wars, working lives, made structures, ways of life and ecologies. Steel—that is the material, the industry and its products—has not only been, but still is "world-shaping." After concrete, it is the most plentiful manufactured material on the face of the planet. Thus, via steel, it becomes possible to shift general imperatives into the particular and the "to hand," as objects of thought and action. All of this is quite different from that pragmatism that says "let's just get on with the job" of sustainability, for without a far clearer sense and understanding of the unsustainable it is not possible to distinguish between symptom and cause, informed action and hollow gesture, or therapeutic versus transformative action.

Another reason why a purely technical account of iron and steel could not adequately convey the actual power of these materials is that iron and steel have enormous symbolic force. Any attempt to think through strategies for reducing the environmental impacts needs to take this symbolic power, this "ecology of meanings," into consideration.

Symbolically, in the Western tradition, steel is the result of Prometheus bringing fire, "the divine spark" of energy and illumination, to earth, whereupon it was adopted by Hephaestus (Vulcan), the god of fire and the forge. And thus, a force of the gods was transferred to the hands of "man."

Iron and steel have been objects of thought, metaphors for power and strength; they have stood for the entirety of the human relation to matter. Consider the view expressed by perhaps the greatest Roman thinker, Pliny, in his *Natural History*:

> It remaineth now, in the next place, to discourse on the mines of iron, a metal which we may well say is both the best and worst implement now used in the world; for with the help of iron we break up and tear into the ground; we plant and plot our groves; we set our vineyards and range our fruitful trees in rows, we prune our vines, and by cutting off the superfluous branches and dead wood, we make them every year look fresh and young again. By means of iron and steel, we build houses, hew quarries, and cut stone; yea, and in one word, we use it to all other necessary uses of this life.

Or consider the philosopher John Locke writing in 1690, some 1600 years after Pliny:

> For it is rational to conclude that, since our faculties are not fitted to penetrate into the internal fabric and real essence of bodies, but plainly discover to us the being of a GOD and the knowledge of ourselves enough to lead us into a full and clear discovery of our duty

and concernment, it will become clear to us, as rational creatures, to employ those faculties we have about what they are most adapted to, and to follow the direction of nature where it seems to point us out the way … Of what consequence the discovery of one's natural body and its properties may be to human life, the whole continent of *America* is a convincing instance: whose ignorance in useful arts and want of the greatest part of the conveniences of life, in a country that abounded with all sorts of natural plenty, I think may be attributed to their ignorance of what was to be found in a very ordinary and despicable stone, I mean the mineral *iron*. And whatever we think of our parts or improvements in this part of the world, where knowledge and plenty seem to vie with each other, yet to anyone that will seriously reflect on it, I suppose it will appear past doubt that, were the use of *iron* lost among us, we should in a few ages be unavoidably reduced to the wants and ignorance of the ancient savages *Americans*, whose natural endowments and provisions come no way short of those of the most flourishing and polite of nations.

More simply, but of the same ilk, here is Harry Scrivenor, a historian of the iron trade, writing in 1854:

It is a doubtful point, whether the domination of man over the animal creation, or his acquiring the useful metals, has contributed most to extend his power.

And it is the fact that this extension of power has been inseparably bound up with the forces of unsustainability that will drive our telling of the stories of the ecologies of steel—a telling absolutely necessary to gain the kind of understandings that can generate conditions of sustainment for futures to be possible.

In the final analysis, we hope that what we present will challenge the thinking of those readers with an existing knowledge of the steel industry by making it possible to view steel from a broader perspective. For those readers who know little about steel, we hope that what follows will not only introduce new knowledge, but also a whole new way of understanding materials, their relation to culture, their place in processes of change and the force they have upon the form of the future.

Structure of the book

Part One introduces the founding moments of iron and steel-making, re-presenting them in ways that can inform the present and future.

Chapter 1 presents a trans-cultural prehistory of iron-making that aims to confound the idea of a linear history and to show how the making of iron was implicated in the development of an "ecology of mind." The spread of knowledge of methods of iron-making from the Middle East to Asia, Africa and Europe evidences the emergence of a traffic in ideas and also demonstrates the "world-shaping" force of ideas. The chapter also examines the advanced iron-making industry of ancient China. Then it looks at the emergence of iron and steel-making in Greek and Roman culture, demonstrating that their methods were far more sophisticated than iron-making at the end of the Dark Ages, which is where most histories of European iron-making start.

Chapter 2 historically reviews the dependence of iron and steel-making on carbon-based fuels (wood, charcoal, coal and coke) explaining thermochemical processes and their environmental impacts. From its inception, iron-making generated environmental problems and there were "environmental crises" from the late Middle Ages. This history is then connected to present-day concerns about greenhouse gas-induced climate change by considering how the steel industry's emissions could be reduced by, for example, newly reinvented charcoal-based methods or by the use of materials like plastic waste as fuel.

Chapter 3 examines how the making of metals and the quest to understand them grew out of a complex collusion between magic, alchemy and metallurgy. The telling of this story of inter-weaving, seemingly incommensurate areas of knowledge runs counter to the more familiar notion of progressive displacement of the one by the other. Magic, alchemy and science continue to coexist in the present as the nature of contemporary advanced materials shows. A case study of a particular alchemist, George Starky, and his connections to Isaac Newton is examined, as is the emergence of process and physical metallurgy.

Part Two considers iron and steel-making as crucial agents of the creation of industrial society. The consequences of the widespread industrial application of iron and steel in war, on the sea and on land, especially in terms of rail and building construction, are major concerns.

Chapter 4, besides looking at the emergent technology of steel-making in the eighteenth and nineteenth centuries and its relation to industrialization, addresses the machine tool industry, specialized tool steel, workplace management and the rise of leadership in engineering from America. The drive to make ever more accurate, precision-performance machine tools and the development of the kinds of steel to make such tools is shown to be pivotal to the rise of industrial mass production and much more. The impact of these developments on the directions the steel industry took is also discussed.

Chapter 5 shows how steel established its presence as the dominant material of the modern epoch by considering some of the major world-forming and transforming applications of steel—specifically modern warfare, railway systems, shipbuilding and the construction industry.

Chapter 6 reviews the "state of the art" of current steel-making technologies in the context of the "state of the world." The fate and environmental implications of integrated steel works, electric arc furnaces and iron substitute materials are considered.

Part Three is framed by the imperative of Sustainment. It confronts the essence of the present and future challenge for the steel industry, which is: given the extent and nature of its environmental impacts, a very significant net reduction of the overall impact of the entire industry is the only way forward. What this means is that improving the environmental performance of the industry while increasing output is just not a viable option. Creating and maintaining a viable steel industry able to advance the ability to sustain, and in so doing create a significant income stream, is thus the pressing challenge.

Chapter 7 argues that the reductive empiricism of environmental science does not have a sufficiently relational picture of impacts and therefore is not capable of dealing with the difficult issue of *structurally inscribed unsustainability*. To counter this, the chapter gives an account of certain environments and ecologies of iron and steel-making at different times and places, attempting to weave together a discussion of biophysical impacts with other impacts less amenable to incorporation by environmental science. The contention is that seeking to understand the *fundamental nature* of the processes of iron and steel-making, and the kinds of environments that they create, is a prerequisite for posing appropriate solutions. The chapter is structured around specific ecologies, or sets of exchange relations within particular environments; it shows how particular materials exchanges such as the extraction, transport and processing of ore and fuel, create distinctive environments which then impact upon other ecologies and environments.

Chapter 8 gives an account of the ways in which the steel industry (and industrial environments more generally) have been sought to be regulated over the last 100 years. It reveals the limitations and contradictions of government control of environmental matters, as well as something more troubling—which is the fundamental limits of current economic and political structures for the advancement of sustainment.

Chapter 9 looks to the future, but not as a vacant space waiting to be filled by utopian visions. Nor is the future viewed with a faith in the ability of science and technology to resolve the mounting planetary problems of unsustainability, as technological determinists believe. The chapter opposes such "future visions" by re-examining the very nature of materials; considering the potential for transformation and redirection by design, design innovation and new standards; addressing the problems of public perception of industry change; and stressing the importance of bringing questions of the immaterial to any new thinking about materials and the economy.

Every effort has been made to ensure that appropriate credit has been given to copyright holders. Any omissions brought to our attention will be addressed in future editions.

Notes

1   The shift was perceived by a number of cultural theorists throughout the
    twentieth century, from Adorno and Horkheimer, who wrote about the
    rise of the Hollywood culture *industry* in the 1930s, to Roland Barthes'
    explorations of the semiotics of mass culture in the 1950s in which he
    noted that it was no longer possible for anything *not* to signify, even
    functionality becoming "the pure sign of functionality," to Jean Baudrillard
    who in the 1960s announced the arrival of the "political economy of the
    sign" in which commodities came to be produced immediately as signs
    and signs as commodities.

2   See, for example, Anne-Marie Willis and Cameron Tonkin *Timber in
    Context: A Guide to Sustainable Use* Sydney: CIS Publications, 1999.

3   Embodied energy is the total of all energy required to make a particular
    material or product (calculated on a per unit basis), including extraction
    of raw materials, processing, manufacturing, transport. For further
    discussion see Bill Lawson *Building Materials, Energy and the Environment*
    Canberra: Royal Australian Institute of Architects, 1996.

4   The most significant material manifestation of an increased velocity
    towards unsustainability was delivered over the course of modernity. See
    Tony Fry *A New Design Philosophy: An Introduction to Defuturing* Sydney:
    UNSW Press 1999.

PART ONE

THE AGES OF IRON

A—Wood.   B—Bricks.   C—Pans.   D—Furnace.   E—Crucible.   F—Pipe.
G—Dipping-pot.

FIGURE 1.1   Metal smelting in Europe as illustrated in Agricola's *De Re Metallica* published in 1556. Source: Georgius Agricola *De Re Metallica* (trans. Herbert Hoover and Lou Hoover) New York: Dover Publication, 1950

1 TRAFFIC IN IDEAS

There are myriad books in many languages across several thousand years that tell of the discovery of the properties of iron, and the subsequent manufacture of iron, then steel. The tale which has been told, retold and elaborated draws upon diverse areas of knowledge: metallurgy, mining, magic, alchemy, chemistry, astrology, philology, economic archaeology and historical anthropology. Reading across these accounts, it becomes evident that much has been missed, errors have been perpetuated, contradictions have accumulated and evolutionary narratives have been allowed to flourish. Overwhelmingly, an instrumentalized history of technique has dominated, obscuring more complex processes of exchange, as well as the geography of an archaeology of knowledge (where knowledge was created and how it travelled).

A distinction needs to be made. There are two histories that touch and part. First, there is the *historicity* of the discovery of iron and of the making of iron and steel (the unwritten historical events across many times and places of human civilization), and then there is the *written history* that gives account of such activity. The former (historicity) is constituted from many actions and events that were separated from each other, often by enormous gaps of space and time. The latter (the *historical literature* that comes to be known as the *history*) imposes order in its writing—selecting evidence, making connections between events, narrating them from a particular point of view. Thus, all history is construction. There is no authentic viewpoint to adopt, nor is it possible to recover every scrap of historical detail. But once the fabricated nature and adopted perspective of any historical narrative is recognized, it becomes possible to raise new questions, revisit primary sources, and strive to read apparently singular narratives as plural and multifaceted.

The recorded history of iron and steel has been dominantly a Eurocentric construction. To our knowledge, there is no general historical account that incorporates both Eastern and Western iron and steel-making. We are not in a position to deliver a fully revised, globally integrated history, and in actuality argue against such "world history" approaches, nor do we wish to eliminate our bias (toward the imperative of sustainment)[1]. Nevertheless, we will be working to correct a number of historical misconceptions.

The literature before printing

In an appendix to the translation of Georgius Agricola's *De Re Metallica*, the translators, Herbert and Lou Hoovers present an historical review of the field of mining and metallurgy.[2] Their account starts with the ancients, moving to the Romans, the medieval, and then to the arrival of the printed text in the West, finishing with the *De Re Metallica* manuscript of 1553 (a little over 100 years after Gutenberg). They consider the Greeks from the −fourth century to the +first century, starting with the considerable influence of Aristotle, then moving to Theophrastus and ending with Dioscorides. Their view is that much of the practical knowledge of the ancient Greeks has been lost because Greek intellectual culture, which created the archive of knowledge with which moderns became familiar, knew little of the skills of practical men. The learned Greeks addressed such matters mostly at a mythological or abstract level—thus there is the account of the coming of metals and fire, delivered by Hephaestus (Vulcan), son of Zeus and Hera, god of fire and the forge, the immortal blacksmith and father of all monsters. While significant because it inscribed the symbolic power of iron, it tells us nothing about the iron-making process. Certainly, Aristotle is credited as having knowledge of iron in his *Meteorologica*. Herbert and Lou Hoovers also draw attention to the Roman scholar, Pliny the Elder, who in his *Natural History* of the +first century, tells of iron being smelted on the island of Elba. Pliny regarded iron as "the best and worst implement now used in the world" and as "… the wickedest invention ever devised,"[3] thus revealing his inchoate sense of the coming of iron as the end of nature and the transformation of war. Certainly, he could look back upon the rise of iron for making of weapons and war machines and see them as superior to those of bronze used by the Assyrians, Persians, Greeks and Romans. The increased sharpness, strength and shock force of an armory of iron axes, knives, swords, arrow and spear heads, generated a corresponding development in shields, chain mail and armor. Newly gained metallurgical knowledge of ores, working of metals and hardening methods also led to rapid advances in the ability to make tools and agricultural implements, while techniques were devised to make larger objects (like ship anchors and heavy chains).

In Europe by the −first century, iron was in increasing use and evident in a range of Roman engineering practices. The most developed were those of the *fabricae*, the craft guild-like college of armorers that was to develop its guarded knowledge over several centuries.[4]

Pliny recognized that iron weapons extended the power of the Roman Empire. Iron was an advanced material that enhanced the Romans' techno-military capability. It should be remembered that wealth at this time was generated through appropriation and conquest. Thus, warfare itself created a demand for iron weapons and implements. One of the main reasons the Romans invaded and occupied Britain was to gain access to high-grade ferrous ores to export back to Rome.

While useful, this Eurocentric account presents only a partial picture.[5] It tells nothing of other regions of the world like, for example, the prehistory and history of iron and steel-making in China. For this, we have to turn to specialist China scholars, most notably the late Joseph Needham, Donald Wagner and the numerous Chinese scholars who supported their research.

There are written Chinese accounts of iron-making from as early as the −fifth century, such as a number of texts providing a commentary on cast iron.[6] Joseph Needham tells us of an ancient calendar, the *Monthly Ordinances of the Chou Dynasty* (the *Yüeh Ling*) which talks of iron as strange among other metals (gold, silver, copper and tin). There is also a chapter of the *Shu Ching* (the *Yü Kung*), a −fourth-century text, which refers to the production of both iron and (cast) steel in an ancient region called Liangchow.[7] By the −third century in China there was a large-scale iron industry. Iron was commonplace (unlike anywhere else in the world) and subject to tax. Cast iron agricultural tools were in use, while iron and quench-hardened cast steel was employed for making weapons.[8] By the first century, metallurgy was under bureaucratic direction (evidenced by the promotion of the first Prefect of Lo-ling [on the plains north of the Yellow River] to the position of "Superintendent of Metallurgical Production.").[9] Archival evidence and scholarship have shown that Chinese thinking and literature on metallurgy was considerably in advance of the Greeks.

According to Donald Wagner, one of the West's most authoritative scholars on the subject, China's iron and steel-making industry was the world's largest and most efficient until about 1700. Its iron and steel industry was on a vast scale with capabilities that remained well ahead of Europe's for many hundreds of years. The Chinese could smelt iron using coal; they built and employed blast furnaces that produced liquid iron in quantity well over a thousand years before such developments in the West. To give a sense of scale, Needham cites the example of the famed ironmaster Cho Shi, who in Szechuan in the –third century founded an iron works that employed nearly 2,000 men.[10]

Conventional accounts present the rise of ferrous materials and technologies in an evolutionary schema wherein small-scale labor-intensive activities considered "primitive" are superseded by large-scale capital intensive organizations with more refined technologies, producing more advanced materials. However, as Donald Wagner notes, "historians of technology are today more willing to accept that technological developments need not run in a straight line from 'primitive' to 'sophisticated.'"[11] The history of the Chinese iron and steel industry is a significant instance of relational, rather than linear progress. Again examples cited by Needham confirm the point—such as the observations of Yang Chhüan, who in his "Discourse on the Principles of Things" (*Wu Li Lun*) of the second century, mentions several metalworkers, including Juan Shih who was said to be highly skilled in the art of fire and water and "the harmony of the hard and the soft"—by which was meant quenching in water to rapidly cool the metal.[12] Quenching was so important that it was a major factor in the siting of iron works. On this Needham refers to a third-century Chinese text which, in recalling practices of the –fourth and –first centuries, discusses the suitability of different types of water (by which was meant, or so it was thought, the effects of varied temperatures, levels of dissolved salts and suspended matter). There was also knowledge of tempering—the practice of altering the quality of metal by slower cooling by immersing hot metal into oils of varied types and viscosity. In contemporary terms, knowledge of quenching and tempering had been developed through empirical trial and error, leading to the realization that the heat of the metal to be quenched and the speed of cooling were critical factors. This knowledge again pre-dated that of European ironmasters by many hundred years.

There is also a lost history that foreshadowed iron-making in the East and West. Its traces have been discerned via interpretive and scientific examination of images and objects, and by philological study of linguistic fragment that reveal knowledge of iron-making in Mesopotamia, Syria, Persia and Egypt. It was in these places that the cultures of iron first came into existence—between 3,500 and 4,500 years ago. One claim is that iron was initially made accidentally while Persians were smelting copper some 4,000 years ago. Prior to this was the use of iron from meteoric sources. Archaeological evidence suggests meteoric iron was used some 6,000 years ago.

Iron and steel-making was advanced in India in ancient times. The Greek historian Herodotus recorded the battle at Thermopylae in the year −480 where Indian archers, using iron-tipped arrows, fought alongside Persians.[13] In contrast to the bronze used by their adversaries, Sparta's military capability in the sixth/seventh century is also claimed to have been based partly on the superior cutting power of their high-quality steel weapons (this has been verified by modern tests of the purity and carbon content of this steel)—although where and how this steel was made is uncertain.[14] There are also accounts of steel being given to Alexander the Great by King Porus in the year −326, which was likely to have been Indian wootz steel (Damascus steel)—a metal celebrated by the Greeks.

The nature of Damascus steel was detailed by Andrew Ure in his influential *Dictionary of Arts, Manufacturing and Mines* (the fourth edition of 1853 gives an account of its contemporary manufacture, which Ure suggests was the same as in ancient times). In summary, "Indian steel or wootz ... consists of the magnetic oxide of iron, united with quartz, in a proportion which do not seem to differ much, being generally about 42 of quartz and 58 of magnetic oxide."[15] The ore was rendered into fine grains by "pounding the ore and winnowing of the stony matrix," the ore and the quartz were mixed with water and packed as lumps into a refractory clay blast furnace (powered by goat-skin bellows) with charcoal and then smelted. After three or four hours, lumps of iron were removed from the base of the furnace, broken up, then the iron and scoria were separated. The resulting very small quantity of iron (less than half a kilogram) was placed in a clay crucible with rice husks and wood chips from selected trees plus some "special" leaves. This was then covered with a clay top and sealed. Once the clay was dry (in about a day), the crucibles were stacked in a furnace and heated under a blast for two hours. When cooled, these crucibles were broken open and "cakes" of steel extracted. These were then turned into bars via a forge and hammering. The bars were taken down in size to thin strips and welded together. They were then taken through the same process of heating and hammering before being worked, for instance, into blades. Wootz steel became famous and known by the trading name of the "emporium" that sold it—"Damascus steel." The renowned strength of this steel resulted from the process of folding and welding, which also gave its surface a distinctive "folded silk" pattern.[16]

Africa: A postscript on the oral

The archaeology and historical anthropology of African iron-making evidences another way that traditional knowledge travelled over time.[17] However, it is not easy to gain an understanding of the contribution of Africa to metallurgy because knowledge of the craft was mostly transmitted by oral culture. The absence of a literature means that the historical picture has been formed mostly by reading of archaeological materials and ethnographic study of popular memory (traditional stories inscribe accounts of knowledge and practice that get passed from generation to generation). It is hard for us moderns to grasp that some of the ancient traditions of iron-making have been passed down through the centuries to the living memory of some older Africans (as anthropological accounts have documented). Equally, some of the early documented histories of iron-making in Africa refer to "lost knowledge of the oral tradition."[18]

There were many centers of iron-making in Africa, some of which continued to produce for perhaps 900 years. One of these was in an area now known as Meroe, in the Sudan, which was a major site of iron smelting.[19] The products of the early industry included weapons (especially swords and lances), chain mail, axes and agricultural equipment,[20] with many such artefacts acquiring considerable symbolic power in the culture. The technology developed in this period, thought to be influenced by contact with Arab cultures, was often well in advance of early European small furnace iron-making that arrived hundreds of years later. It was common for African furnaces to be broken down after each smelt, to use a hot air blast to increase furnace temperature and to manufacture charcoal of different qualities to vary the properties of the iron produced—which meant that early Africans had the ability to vary the carbon content of the iron (which is the basis of the claim that steel was made in Africa in ancient times).[21]

In the view of archaeologists working in Africa, iron-making started in the Sahel region of the north and then moved toward the west and southeast. In modern geographic, colonially inflected terms, the Congo, Rwanda, Burundi, Zaire, Mali, Malawi and Zimbabwe were all significant iron-making areas.[22] Iron-making occupied a position of considerable status in Africa, although the blacksmith was a figure of varied significance in many of its cultures. In Rwanda, for example, iron-making played a very important and positive role in the symbolic life of the culture, whereas in Ethiopia blacksmiths formed a despised caste who were barred from owning land, cattle and participating in ceremonial activities.[23]

Design, things and the traffic in ideas

Design history is silent on the design of ancient technologies and artefacts. Yet obviously, such "things" had a determinant presence. While archaeology makes this clear, recovering information on the dissemination of, and trade in, knowledge and ideas is nonetheless both imprecise and complex.

Oral traditions, the passage of goods, knowledge of their designed form and method of manufacture (together with the circulation and translation of texts, war and conquest, education, accident and the agency of images) all played their part in transmitting knowledge, often over an extended period of time. The example of the geo-cultural passage of classical Greek thought is illuminating. It took hundreds of years for the classics to arrive in the English language and its culture. The Greek texts were first translated into Arabic, then into Latin, then finally into English and other languages. Time and translation clearly modified the meaning of original texts. There are also various accounts of the traffic in ideas by early traders, as well as of artefacts, especially by the Phoenicians. All of this meant that the movement of manuscripts, and the language they used, transformed meanings—especially of how an object might be understood and represented when moving from one culture to another.[24]

Our concern here is not about whose history is true or false, or who led in the development of materials and techniques, or even whose material attainments were employed to help establish cultural superiority. Rather it is with what, in 1973, Gregory Bateson called an "ecology of mind"—a notion resting on an "ecology of ideas." He defined this through "a way of thinking" (outlined in a collection of essays written over 35 years), saying:

> It is a science which does not yet exist as an organized body of theory or knowledge. But the definition of an "idea" which the essays combine to propose is much wider and more formal than is conventional. The essays must speak for themselves, but here at the beginning let me state my belief that such matters as bilateral symmetry of an animal, the patterned arrangement of leaves of a plant, the escalation of an armaments race, the processes of courtship, the nature of play, the grammar of a sentence, the mystery of biological evolution, and the contemporary crisis in man's relationship to his environment, can only be understood in terms of such an ecology of ideas I propose.[25]

Bateson's neo-Kantian claim is that ideas are ecological—they are directive, they objectify, they make things happen.

An "ecology of mind" can be understood as a collective thinking and memory formed out of the unorganized processes of the world in action. The crucial point is that "mind" is not simply internal to the body, but is present in multiple forms—it is inscribed in making, using and inhabiting the world. Mind so inscribed animates made things, the processes of making, material structures, modes of organization, and of course, acts and forms of communication and exchange, both written and spoken. Extending Bateson's idea, we can say that an "ecology of steel" is closely interconnected with an "ecology of mind." This is to say that iron and steel-making as an appropriation and transformation of raw materials, depended upon a traffic in ideas, invention, discovery and the application of processes that were informed by particular ecologies of mind. At the same time, such activities acted back upon the formation and transformations of those ecologies of mind.

The transmission of ideas and the trade in designed objects, underpinning an "ecology of mind" was sometimes fast, but mostly very slow. Objects in transit were just as much at the mercy of interpretation as were ideas on the move. The nature of a "thing"—its qualities, function, form, matter and its inscribed fate (the value posited with it that determined its "life" as anywhere between the poles of the precious and the ephemeral)—was determined according to the interpretive act in its context. It is now "natural" for most of us, for example, to ask questions of some "thing" like: What is this? How does it work? What is it made from? Who designed it? Why was it made? How was it made? Such questions and their answers are the product of particular cultural conventions which constitute the object in an instrumentalized materiality centered on ideas in action. However, such seemingly neutral questions are in fact the product of a particular "ecology of mind." Dramatically different, other ways can be brought to the thinking of some "thing" such as: Where is its source? Is it sacred? Whose magic brought it into being? Or—will it steal my spirit? Out of everyday and often almost unnoticed encounters with things, knowledge is transmitted, extracted or invented. Knowledge, as a product of an "ecology of mind," arrives out of dynamic and often chaotic circumstances, rather than being based upon an orderly and invariable correspondence between representational form and idea. Ideas and actions emerge with an energetic force that animates them—just consider the massive transformation of loaded meaning over time and culture of seemingly simple things: a rock, tree, fire, earth, water. Conversely, much withers and dies because of a lack of receptivity.

Textual materials, books, histories of relevant topics do not have to have been produced with cognizance of the ecology of mind nor subscribe to its theory of knowledge, to be a part of it.

Historically, the ability of oral culture to spread knowledge was limited by the capacity of memory (be it that memory in oral cultures was highly developed). It was not until the arrival of the written word, and an appropriate substrate for writing and then printing, that the kind of traffic in knowledge and ideas which gave rise to a modern ecology of mind became possible. The speed and spread of knowledge thereafter was determined in other ways, like comparative levels of literacy, archival record-keeping, the textual production of an intellectual strata of society, the output of a printing industry and the value given to concrete and abstract knowledge by a particular society. The development of writing, printing and literacy was very uneven between nations—those that were the most advanced in these areas acquired greater agency to transmit knowledge.

Taking up the idea of "ecology of mind," we will now consider one of the formative events of the European literature on metallurgy, followed by discussion of the ideas inscribed in the first practices of Western iron-making. We will then revisit China to see a different ecology of mind that was operative in its early literature and iron-making practices.

The founding moments

As already noted, European texts with information on metallurgy were not produced until after a century of the arrival of printing. By this time, some 30,000 books had been printed (mostly theological works and the classics).

In these early texts, information about metals, mining, ores, metallurgy, alchemy and chemistry was all mingled together. This was because such knowledge was not yet systematically subdivided into the separate disciplines. Asian history tells another tale. First of all, printing with movable blocks had been invented in the ninth century, some 500 years before Gutenberg, and secondly, one of the earliest printed texts, the "Bibliography of Lung Hung," to which Needham draws to our attention, detailed practical experiments in alchemy, metallurgy and chemistry.[26]

The word of Europe

The two founding texts of European metallurgical literature, arriving within 20 years of each other, challenge any simple notion of a unified European sensibility. They are symptomatic of quite different ecologies of mind.

The *Pirotechnia* of Vannoccio Biringuccio is regarded as the first printed work in the West to "cover the whole field of metallurgy."[27] The text was printed a year before the death of its author in Venice in 1540. It consists of ten books divided into short chapters starting with an account of eight ores, followed by a review of minerals. Books 3 to 6 cover assaying, the separation of gold from silver, alloys and casting. Book 7 addresses methods of melting metals, including the description of a reverberatory furnace for making bronze—this well over 100 years before such a furnace was meant to have been "invented" for smelting non-ferrous metals.[28] Book 8 deals in more detail with the process of casting (whereas Book 6 was mostly concerned with the technology required). Book 9 deals with "works of fire," including the management of slag. Finally, Book 10 addresses combustible materials. Two immediate impressions are gained from even the most cursory of readings. The first is that it is a remarkable work of empirical enquiry by a craftsman scientist. The second, is just how restricted was the traffic in ideas (the "ecology of mind") of iron-making amid all the other crafts, arts and sciences. In an age in which the typographic word was dominantly the currency of a cloistered culture of learning, knowledge of practical matters travelled very slowly and usually within very limited communities of practice. Biringuccio's remarkable book of applied practical inquiry and proto-technical literature (of which there were very few) was "disadvantaged" by being written in vernacular Italian in an age when Latin was the language of learning.[29]

Chapter 17 of Book 1 of the *Pirotechnia* concerns Biringuccio's method of making steel—which he understands as "… nothing other than iron, well purified by means of art and given a more perfect elemental mixture and quality by the great decoction of the fire than it had before." The steel he refers to is what we now know as "crucible steel," a cast material often presented as an intermediate material between early methods of making iron and modern steel-making (there are similarities here to the Chinese methods of making cast steel, which, as we shall see later, confounds this chronology). Biringuccio's method consisted of immersing wrought iron "blooms" (low carbon elemental iron from which a significant amount of slag, but not all, had been removed by the forge hammer)—weighing something of the order of 18 to 20 kilograms—in a bath of cast iron (effectively re-smelted "pig iron"—that is, iron reduced from ore in the furnace that still contains all its impurities, including a large amount of carbon). The process was continued, at a constant heat, for a period of six hours "often stirring with a stick as cooks stir food."[30]

Chapter 3 of Book 3 gives us a picture of the scale of a blast furnace:

I wish to tell you how the means that are used for smelting and purifying iron are really blast furnaces, although they are just called furnaces. It is indeed true that they are larger and arranged in a different manner from ordinary ones because a larger amount and a greater violence of fire is required on account of its poorly mixed earthiness. For this reason, those large bellows and those large interior spaces for holding charcoal are made. I have seen some of these blast furnaces eight *braccia* high and two and a half wide in diameter at the centre and two at the bottom. Whoever wishes to make this well should cut it in a hillside so that the ore and charcoal can be put easily from above on level ground, which bears the load that the animal brings there, for the blast furnaces are never so small that they need less than fifty or sixty sacks of charcoal and six or eight loads of ore continually. Moreover, it is not surprising that it is also necessary to have large bellows since much blast is necessary to keep such fire alive.[31]

On this description, we note that the ore used to charge the furnace was not carelessly loaded but preceded by a rigorous process:

... whoever wishes to make iron good and soft by virtue of the ore itself ... must first provide an experienced and intelligent sorter who will carefully sort the pure from the impure and will separate them according to the indications of their appearance and by breaking them. Then he will roast them in an open furnace and, thus roasted, he will put them in an open place so that the rains will wet and the sun will dry them out. Having left them thus for some time, he must look them over again piece by piece before they are brought to the furnace ...[32]

While Biringuccio, as would be expected, presents charcoal, as well as wood, as the "the food of the fire," he also, again contrary to conventional history of iron-making in Europe, makes a somewhat oblique reference to the use of coal:

Besides trees, stones that occur in many places that have the nature of true charcoal; with these the inhabitants of the district work iron and smelt other metals and prepare other stones for making lime for building.[33]

As with the example of early Chinese knowledge of quenching and tempering, Biringuccio confounds the historical ordering of knowledge as a single line of development. In writing of the hardening of iron and steel, he recognized that secret knowledge was an inherent feature of the "ecology of mind" of the craft of making and working metals. (Later histories of the practices just did not capture this knowledge nor notice its absence.) Biringuccio identified, but did not fully reveal, such secrets, including those on quenching and tempering:

... the various temperings with water, herb juices or oils, as well as the tempering of files. In these things, as well as in common water, it is necessary to understand well the colours that are shown and thrown off on cooling. It is necessary to know how to provide that they acquire these colours well in cooling, according to the work and also the fineness of the steel.[34]

Although Biringuccio was not without some education in the classics, his *Pirotechnia* was an exercise of "practical reason," directed towards metal workers. Its place was on the workshop shelf for reference rather than in a library. In contrast, the *De Re Metallica* of Agricola, completed in 1550, but not printed until 1556 (a year after his death), was a work of learning in its style, content and appearance. It was a nodal point in quite a different "ecology of mind." The book was destined to be elevated as a highly designed object of display in the libraries of the cultivated (and it was). Notwithstanding the difference between these two books, there was an unambiguous relation between the ecologies they engaged. In part, the tenor of this relation is indicated by Agricola's perhaps contestable remarks on Biringuccio in his Preface, when he says:

... a wise man, who wrote in vernacular Italian on the subject of the melting, separating, and alloying of metals. He touched briefly on the methods of smelting certain ores, and explained more fully the methods of making certain juices, by reading his directions, I have refreshed my memory of those things which I myself saw in Italy; as for many matters on which I write, he did not touch upon them at all, or touched but lightly.[35]

To these comments are added the views of Agricola's translators, Herbert and Lou Hoover, who assess Biringuccio's work thus: "his descriptions are far inferior to Agricola's; they do not compass anything like the same range of metallurgy. And betray the lack of a logical mind."[36] These dismissive remarks obscure the fact that Agricola and Biringuccio were attempting to deliver almost identical projects, one from a position of privilege, the other from a condition of labor. Metallurgical concerns occupy almost the totality of Biringuccio's text, whereas the first seven of Agricola's 12 books are concerned with mining and what the Greek and Roman classics had to say on metals. Besides his books on assaying, ores, smelting, the separation of metals and juices (liquid and solidified

A—AXLES. B—WHEEL WHICH IS TURNED BY TREADING. C—TOOTHED WHEEL.
D—DRUM MADE OF RUNDLES. E—DRUM TO WHICH ARE FIXED IRON CLAMPS.
F—SECOND WHEEL. G—BALLS.

FIGURE 1.2 Waterwheel from *De Re Metallica* published in 1556

minerals) Agricola addresses the knowledge required by miners of topography, water, veins of ore, surveying, methods of measuring, tools, mining machines and techniques in mining technology. So said, it is worth qualifying Agricola's acknowledgement of a debt to Biringuccio with what Smith, Biringuccio's translator, tells us (his remarks also provide a comment on the view of the Hoovers):

> Agricola's "refreshing of memory" consisted of copying *in extenso*, without further acknowledgement, the earlier author's account of mercury and sulfur distillation, glass and steel-making, and the recovery by crystallisation of saltpetre, alum, salt and vitriol together with other less important sections. Agricola usually added a superior illustration and often provided valuable additional detail.[37]

A comparative reading of both texts on topics like ore, smelting and furnace technology confirms the fairness and accuracy of these remarks.

Obviously both books have become of enormous historical interest, and certainly Agricola's text, and its numerous and oft-used illustrations, have long since acquired a great deal of informational significance which completely transcends the consequences of its original aestheticization of its topic.

The words of Asia

The intercultural spread of iron-making knowledge beyond its first makers cannot be separated from the means of communication by which this knowledge travelled. The historical significance of China needs very much to be seen in this light, not least because of a convergence between its early attainments in the craft of iron-making and the related significance of its highly developed culture of documentation, learned manuscript production and early development of printing. However, for a very long time, the texts circulated *almost* exclusively within China, or within closed communities elsewhere who guarded their secret knowledge. These circumstances were obviously a major delimitation of the global influence of early Chinese civilizations.

It is only now, as a result of scholarship (frequently linked to the collection of archival material by academic institutions outside China), that some of the information contained in these ancient texts has started to become more widely known in both China and the West. These texts reveal the degree of advancement of ancient Chinese iron and steel-making, prompting a requirement for a substantial revision of the (Western) history of the field—which is slow in coming.

As indicated, the Chinese developed the ability to smelt iron to a liquid state very early in their history of iron-making. Unlike elsewhere, Chinese iron-making was not preceded by an era of bloomery production (an early method of making iron in the West in which the metal never became fully liquid, because sufficient temperatures could not be reached; instead the iron in the ore was formed into a pasty ball to be worked by a forge hammer into a usable wrought material). The Chinese also gained the ability to use coal as a fuel almost as soon as they gained knowledge of how to make iron. One method was to partly fill crucibles of good refractory clay with crushed iron and anthracite (coal) which were then placed on a bed of anthracite and subjected to a continuous blast of cold air. Chinese iron-making also had the advantage of good grade ores, ready-to-hand excellent refractory clays (to line crucibles and furnaces) and plentiful supplies of coal—these factors, combined with the ability to deliver a constant blast from blowing engines, enabled high temperatures to be reached and held. They also learned, very early, how to deal with sulfur contamination (a problem that plagued attempts to make iron with coal in the West for a very long time) by the addition of limestone with which it fluxed and then was held. This process produced iron that could be cast or used as a hot metal for further refining. These techniques were being employed in China well over 1,000 years before the early modern iron workers of the West were able to produce liquid iron.[38] The success of the Chinese was due to the power of their bellows and blowing engines— the latter being able to deliver a continuous blast at a very early stage of the technology, initially by the double action of their piston bellows.

The ability of the Chinese to create high furnace temperatures had a number of major implications: it meant that "pig iron" (iron with a carbon content of 1.5–4.5 percent) could be produced and cast (the same material in China was called "raw iron"). Basically, pig iron resulted from smelting iron ore with a carbon source—initially from charcoal, then certain grades of very low sulfur coal, and then coke. Smelting methods improved by the introduction of a mineral flux to create an initial composite of iron, carbon and slag. The carbon source in the blast furnace had two functions: a metallurgical function in the thermochemical process as a reductant; and in the thermal process as a fuel. However as we have seen, depending on the method of iron-making, these processes could be combined in the charge fed to a blast furnace or were operated indirectly as in a crucible method where the thermochemical and thermal functions were separated from each other.

To give some further sense of the comparatively advanced organization of iron-making in China, we note that by the second century there was an imperial China "Iron Casting Bureau."

Cast iron is hard, non-malleable, extremely brittle and unable to be welded.[39] It was used for many artefacts that were not subject to impact (like cooking pots). Besides its carbon content, cast iron contains many impurities, especially sulfur, silicon and phosphorus, which increase the melting temperature. Their removal required oxidization (extreme heat plus oxygen), which then produces "pure iron," which the West called wrought iron (but the Chinese called "ripe iron"). This material has the opposite qualities of cast iron—it can withstand impacts, is malleable, although soft it is tough, but is unable to be hardened. Wrought iron has very low carbon content (of the order of 0.05 percent—which is why it is soft and ductile). Its strength was increased by the presence of slag, giving it a fibrous quality and in actuality making it a composite material.[40] Over an extensive period of time, the applications of wrought iron increased. In the nineteenth century, prior to its displacement by steel, it was used, for example, for things as diverse as water pipes, nails, ships' hulls and horseshoes—it has some remarkable qualities (not least, resistance to corrosion in salt water).[41]

Steel, at its most basic, was a material created to combine the desired qualities of the hardness of cast iron with the malleability and workable qualities of wrought iron.[42] There is, of course, a difference between early steel, coming from inexact "trial and error" processes, and contemporary steel produced according to modern metallurgy, which delivers a highly controlled material with diverse performance characteristics.[43] Initially, the desired performative qualities of a metal were the result of practical experimentation, rather than application of conscious knowledge about combining two elements, iron and carbon, to create an alloy. Even a small variation in the amount of carbon changes the crystalline structure, and thus the performative characteristics, of the metal. Early steel-making like the Indian "wootz" method (as outlined by Ure) and the Chinese method of co-fusion (which produced a cast steel) are both examples of empirical "solutions" to localized performative demands.[44]

The form of steel-making that Biringuccio and Agricola were aware of consisted of submerging pre-heated lumps of "pasty" wrought iron in molten cast iron to create surface fusion between the irons, extracting it and working it by cycles of forge hammering and reheating to de-slag it and to evenly distribute the carbon. Not only did such an imprecise method not deliver steel with a controlled carbon content and low level of impurities, it resulted in markedly inferior steel than that made in the Middle East and Asia in much earlier times.

As already indicated, the Chinese were early steel-makers—their steel being made by adding carbon to wrought iron or extracting carbon from pig iron. In the fifth century the Chinese were also using the "co-fusion method." Here, wrought iron (high melting point 1535°C) was immersed in a bath of cast iron (low melting point 1130°C), completely melted and mixed. A measured ratio of cast and wrought iron was placed in a cupola and taken to a high temperature over the hearth of a furnace for several days and nights. The steel was then cast. The Chinese were able to reach the higher temperatures required to perfect the system earlier than European ironmasters and steel-makers because of the blowing machines they invented. During the same period the ancient Chinese produced cast iron in volume and developed techniques to decarburize cast iron to make steel by a process they called "hundred refinings." This used controlled blasts of cold air to oxidize carbon and impurities. This method of steel-making was, as Needham points out, "theoretically ancestral to Bessemer conversion." He also observes, very interestingly, "direct migration of Chinese workman skilled in this work immediately preceded the group of inventions associated with the name Bessemer." This is not only another indicator of the traffic in ideas, but indicates the need for an extensive historical inquiry into Western culture's appropriation of Eastern knowledge and material practices, from the Enlightenment onward.[45]

Side by side with the co-fusion process, the Chinese also employed a refining process to transform pig iron to wrought iron. The perfection of this process in the West had to wait until the arrival of the puddling method patented by Henry Cort in 1784 (a process based on a reverberatory furnace).[46] Cort's furnace operated by a flame from a coal-fired hearth being drawn over and reflected down onto a crucible of molten iron with gases extracted on the opposite side of the crucible to the hearth.[47] Pig iron was melted in the crucible and, when molten, stirred with a rod (hence "puddled"), this constantly exposed the surface to flame and oxygen which oxidized the carbon and created a puddle ball, which was lifted out with tongs and forge hammered into wrought iron.

Cort's furnace was displaced by the more sophisticated design of Siemens-Martin steel-making process in the 1860s—a high temperature regenerative furnace with an acid slag lining able to complete large volume steel-making in the order of 16 hours.[48] The Siemens-Martin process reconnects us to co-fusion.

William Siemens (a German resident in England), with his brothers, was a builder who advanced the performance of open-hearth regenerative furnaces from the mid-nineteenth century. By 1867, he was making a high temperature reverberatory furnace which could melt pig iron and burn off carbon, silicon and manganese with an oxidizing flame. Although the melt temperature increased as the carbon was reduced, the furnace had the ability to keep the metal liquid—it thus could out-perform the puddling furnace and make steel. Around the same time, two French furnace builders, Emile and Pierre Martin, conceived and built an open-hearth furnace in which it was possible to add wrought iron (including scrap wrought iron) to molten pig iron. Here then was co-fusion writ large. The combination of the Siemens design and the Martin steel-making methods, with the addition of the dolomite (carbonate of lime) lining (an 1883 innovation by two Welshmen, Thomas and Gilchrist, which caused the slag to retain phosphorus and thus prevented it contaminating the steel) made it possible to use the large available amounts of ore with phosphorus content for steel-making (this had enormous consequences in the USA—where a great deal of ore was high in phosphorus, a fact that inhibited the development of the industry in the USA). The Siemens-Martin process with the Thomas-Gilchrist innovation established the basis of the modern open-hearth furnace and steel-making process.[49]

A significant Western method of making cast steel, with some similarities to Chinese co-fusion, was perfected by Benjamin Huntsman in 1742 in Sheffield. Huntsman used "blister steel," a material that was produced in a cementation furnace by wrought iron bars being packed into a stone box, with a carbon source (charcoal dust), sealed with sand and clay, and heated to a bright red heat for about five days. This baking defused carbon into the surface of the wrought iron, and in so doing the carbon reacted with the iron oxide in the slag and released carbon monoxide, which in turn produced the surface blistering—hence the name. This work was hot, hard and extremely dirty! The basis of the Huntsman method was to melt cut bars of "blister steel" in a crucible, thereby homogenizing the low carbon "sap" iron with the carburized steel that surrounded it. This hot metal was then cast into ingots. All of this required a great deal of skill: the timing, temperature and chemical process were all matter of critical judgment exercised by craft workers—there were no scientific measuring instruments to hand! The combination of the process with the high-grade iron used for cementation (a low sulfur, low slag charcoal-smelted iron imported from Sweden) resulted in the highest quality of the best steel available in its day.

Blister steel was also used to make "shear steel"—a method of steel-making close to Indian wootz steel. Shear steel was made by welding strips of blister steel together and then forge hammering to work it into a more homogenized material. It was this steel that made the city of Sheffield a world leader in tool and cutlery manufacture for nearly 200 years.

While the foregoing account is a long way from being a detailed picture of the complexity and diversity of steel-making, it does reinforce the view that the development of the technology of iron and steel-making did not proceed along a sequentially ordered evolutionary path. But the disjunctural development between East and West still needs further elaboration.

Needham brings to our notice a text authored by Sung Ying-Hsing in 1637, the *Thien Kung Khai Wu* (translated as: "The Exploitation of the Works of Nature"). Among other things, this text gives an account of the Chinese blast furnaces—this at almost exactly the same moment as Biringuccio was writing his *Pirotechnia*.[50] We should remember that the ability to produce cast iron in China had existed from the –fourth or –fifth centuries. As already discussed, the Chinese cupola method of making cast iron was not pre-dated by bloomery iron as was the case in Europe.[51] Moreover, while Biringuccio describes a proto co-fusion process, as we have just pointed out this method of steel-making had already been practiced in China for over 1,000 years.[52] Small blast furnaces were developed early in the history of iron production in China, and by the time of Sung's text, wrought iron was being made by furnaces with large double action bellows and puddling hearths into which the iron flowed after tapping. In this process, besides stirring the molten metal to assist oxidization, silica was added to assist decarburization. Again, as mentioned, one of the most important innovations of Chinese iron-making was the power of their bellows and blowing engines, which were able to deliver a continuous blast by the double action of their piston bellows.[53] Needham suggests this technology may have existed from, at the latest, +third century or perhaps as far back as the −fourth century; additionally, water power (via the water wheel) was used to drive bellows from about the +first century.[54]

It is even more remarkable that from very early times the Chinese were also making steel by "the tricky process of direct decarburization of cast iron under a cold blast."[55] One of the "tricks" in this process was to add iron oxide (which, as an oxygen-donator, assisted in oxidizing carbon away).

As said, the Chinese had the ability to make cast iron and steel in great abundance long before the Europeans, but the socioeconomic circumstances of its production (including the prevailing ecology of mind) were very different from those in Europe at the beginning of its industrial era. As Needham and others have pointed out, the production of iron and steel in China was not constrained by the capability of the technology but by the limited market demand of a pre-industrial society and by the restrictive character of the feudal-bureaucratic structure of the society. This view is supported by a wealth of archaeo-historical evidence.

The material attainments of the Chinese industry are to be seen in many forms, such as cast iron roof tiles, plaques, chains, a 13-storey-high cast iron pagoda (of the tenth century) and an iron chain bridge (built 1,000 years before Europe's first suspension bridge in the sixteenth century).[56] Likewise, the generalized use of iron ploughshares and agricultural tools transformed the "nature" and productivity of farming in China, putting it in advance of European methods by centuries. Iron and steel also played a major part in arming Chinese armies.[57] Finally, and perhaps most remarkable of all, we need to wonder at the ability of Chinese craft workers to experiment, invent and manage complex processes without any conceptual picture of chemical processes or theoretical knowledge of metallurgy.

As we have been saying, the history of iron and steel-making was not discrete. The knowledge of materials and techniques existed in a geographically dispersed "ecology of mind" that had backward and forward movements and traces in many parts of the world, especially Persia, India, Turkey, Greece, China and Africa. The historical importance of China was recognized in the West a very long time ago by only a few perceptive individuals. Scrivenor's comments, in his history of the "iron trade" of 1854, are telling and poignantly gel with the work of contemporary Western scholars like Needham and Wagner. He writes:[58]

The literary records of China may, when explored, open, at some future date, many interesting facts in the history of Scythians, Tartars, and Russians: their early trading, and extent of their knowledge. It is much to be lamented that as vast a field of antiquity as China has not yet been fully explored and ably gleaned. In that isolated empire, the arts and sciences flourished for ages anterior to the era of our Lord.

A few lines later, he then says:

This brief notice, we conceive, justifies our expectation that China possesses much which we think may enlighten the industrious inquirer.

While there is no doubt an archaeology, chronology and geography of invention that was textually inscribed and communicated by print, a great deal of the substance of oral culture and memorized histories was erased. Most of what we know of early periods is based on those textual sources that survived—well beyond the significance of the history of iron and steel, they play a major role in creating a picture of an earlier ecology of mind. An enormous amount of the knowledge and skill of craft workers and their varied cultures simply disappeared. Yet this knowledge and practice constituted a very significant part of the historicity of iron and steel-making. There is always significant difference between what makes up the actual forces and features of a historical moment and the retrospective telling of a history as it is written and printed—this is the fundamental difference between *historicity* (actual events) and *history* (the selection, editing and narrativization of events).

A China postscript

Mass production, a factory system and work forces of thousands were all elemental to Chinese economic activity well over 2,500 years ago.[59] Methods of production were often documented in design manuals, many of which were not rediscovered until the mid-twentieth century.[60] These documents provide a great deal of information on a key principle of interest to us—the principle of modularity. It is in fact at the very core of Chinese culture and was the basis of its written script. Building, pottery, armies, artworks, printing, bronze casting, and much more, were organized on this principle.

One of the manuals with particular relevance to the contemporary imperative for sustainable construction was the *Yingzao Fashi*. This famous and influential manual was written in 1091, with a second edition in 1103 (no evidence of the first edition still exists). It was a design and technical manual of standards for the Master of Works, which was a section in the Ministry of Works, the government department responsible for the construction of palaces, temples, barracks, government buildings, moats, gardens, bridges and boats.[61] It was produced to deal with the massive expansion of building development in the first 100 years of the Sun Dynasty.

The second edition of the *Yingzao Fashi* addressed the ordering of materials, building design, construction details for all building types including the detailing of stonework, carpentry and joinery, wood carving, roofing, plastering and finishes. One of the key features of the manual was its use of a modular standard of measurement (a *fen*) that in many ways prefigured systems building.[62] What is remarkable is the way the modular design methods allowed for a new building to employ components taken from the disassembly of an old building of a different scale and use.[63] In modern terms, what it provided were instructions for adaptive reuse. While having enormous status as a document in Chinese architectural history, the significance of the *Yingzao Fashi* to contemporary design practice, in and beyond architecture, has not yet been comprehended.

Read from the perspective of today's imperatives, it provided instruction on design for the conservation of materials; waste elimination; design-for-disassembly and movable buildings; interchangeable components; and, above all, the value of a design-based tradition of construction standards.

One can contrast this ancient thinking with today's, such as the contemporary contradiction of attempting to load "environmental performance" onto individually expressive and aesthetically overcooked buildings that often have a restricted design-life. Rather than demonstrating "creativity" this design disposition displays a limited imagination which reduces "the designed" to technology and mere appearance. The search for "another imagination" is hardly yet a glimmer on the distant horizon.

During the second half of the twentieth century, China moved from being a very minor steel-maker to the world's largest producer. The opening of the twenty-first century heralded China's second coming as a globally dominant industrial force. This is not just based on sheer quantity of industrial output, but also on rapid improvement of quality. The big issue in this situation is not so much whether the position of dominance will come to pass, but rather, the model of leadership that gets established.

The volume of China's steel production is being driven by domestic demand from its construction industry trying to meet the needs of rapid urbanization and from its manufacturing sector (these two markets represent over three-quarters of China's total domestic steel consumption). However, China is also importing steel because it is unable meet the highest quality requirement for certain sectors of the economy (in particular, the fast-growing auto industry and to a lesser extent, the electrical appliance industry). To rectify this, a major investment and technology upgrade program has been ongoing. On this count, China's development disposition is unquestionably extremely problematic. All the signs are that the development vision of "catch-up" is still driving the national economy—a vision based on the unsustainable economic and cultural history of modernity. This backward-looking vision, resting on an outmoded image of "progress," is linked also to political and cultural devaluation, as well as erasure of traditions of the recent and distant past.

Leadership in the unfolding epoch cannot be predicated upon extending the current paradigm which inscribes the error of the unsustainable. It has to move to the next one—the paradigm of "sustainment," that is nevertheless fully aware of the problems of paternalism, neo-colonialism, double standards, but is committed to the practical necessity of redistributive justice. On this count, there is more to learn from China's past than its present.

Notes

1   The term sustainability has a number of frequently problematic contemporary usages, not least its ambiguous association with "ecologically sustainable development"—where it encompasses both the ecological and the economic (as development). In order to make a distinction from these usages we have adopted the term "sustainment" throughout this text. The implications of this shift in terminology has

been developed in Tony Fry "The Sustainment and its Dialectic" *Design Philosophy Papers Collection One* Ravensbourne (Qld, Australia): Team D/E/S Publications, 2004.

2   Georgius Agricola *De Re Metallica* (trans. Herbert Hoover and Lou Hoover) New York: Dover Publication, 1950, Appendix B. This text was originally translated in 1912.

3   Harry Scrivenor in *History of The Iron Trade: From the Earliest Records to the Present* London: Longman, Brown, Green and Longmans, 1854, makes substantial reference to Pliny, pp. 13–18.

4   There is no developed history of the iron industry in Britain after the departure of the Romans and during the Dark Ages (to a large extent because of a lack of historical record-keepers besides the few key documents of church and state, such as the Domesday book). It is not until the eleventh and twelfth centuries that records start to become more common. By the time of the Crusades, records of orders and accounts for items like weapons, armor plate and chain mail begin to appear. Aitchinson, for example, cites an order for 50,000 horseshoes and points out that a suit of armor cost the equivalent of a modern tank. Leslie Aitchison *A History of Metals* (Volumes 1 and 2) London: MacDonald and Evans, 1960, Volume 1, p. 334.

5   The technical literature of which, as the Hoovers say "… could be reproduced on less than twenty of these pages," Georgius Agricola *De Re Metallica* (trans. Herbert Hoover and Lou Hoover) New York: Dover Publication, 1950, Appendix B, p. 607.

6   Joseph Needham *The Development of Iron and Steel Technology in China: Second Biennial Dickinson Memorial Lecture* London: Newcomen Society, 1956, p. 5.

7   Ibid.

8   Ibid., pp. 5–6 and Joseph Needham *Science and Civilisation in China Volume 5: Chemistry and Chemical Technology,* Part 2: "Ferrous Metallurgy" (by Donald Wagner) Cambridge: Cambridge University Press, 2008, p. 130.

9   Joseph Needham *Science and Civilisation in China Volume 4: Physics and Physical Technology* Cambridge: Cambridge University Press, 1965, p. 370.

10   Joseph Needham *The Development of Iron and Steel Technology in China*, p. 7.

11   Donald B. Wagner *The Traditional Chinese Iron Industry and its Modern Fate* Copenhagen: Nordic Institute of Asian Studies, 1997 (web version), p. 4/60.

12   Needham *The Development of Iron and Steel Technology in China*, pp. 24–5.

13   See U. S. Yadav and B. D. Pandey "The historical perspectives of Indian iron-making" *Steel Times* April 1999, p. 145.

14 Douglas A. Fisher *The Epic of Steel* New York: Harper and Row, 1963, pp. 23–4.

15 Andrew Ure *Dictionary of Arts, Manufacturing and Mines* London: Longman, Brown, Green and Longman, 1853, pp. 731–2.

16 As indicated, "Damascus" was the name of the emporium that sold this steel rather than the place where it was made.

17 See Randi Haaland and Peter Schinnie (eds) *African Iron Working: Ancient and Traditional* Oslo: Norwegian University Press, 1985.

18 See, for example, the influential Harry Scrivenor *History of the Iron Trade*, pp. 19–22. Scrivenor also refers to iron-making in the Himalayas and in pre-colonial Peru where mountain furnaces (designed to smelt silver) were located to capture wind.

19 F. J. Kense "The Initial Defusing of Iron to Africa" in Randi Haaland and Peter Schinnie (eds) *African Iron Working*, p. 21.

20 Haaland and Schinnie (eds) *African Iron Working*, p. 51.

21 Ibid., pp. 54–7.

22 Kense "The Initial Defusing of Iron to Africa," pp. 20–3.

23 Haaland and Schinnie (eds) *African Iron Working*, pp. 73–87, p. 91.

24 This process touches our story—Joseph Needham cites an example from the work of Haudricourt, who presented evidence on the variable characteristics and meanings of "cast iron" as a material and as a word designating a material, as it moved from Asia to the Middle East and then to Europe via Turkey. Needham *The Development of Iron and Steel Technology in China*, p. 22.

25 Gregory Bateson *Steps to an Ecology of Mind* St Albans: Granada/Paladin, 1973, p. 21.

26 Needham op. cit., pp. 167–78.

27 Vannoccio Biringuccio *Pirotechnia* Cambridge, MA: The MIT Press, 1966 (reissue of the 1942 edition).

28 In this type of furnace, air was drawn from fuel burning in a firebox back into the furnace rather than the ore being in direct contact with the fuel. The reverberatory furnace, and its associated puddling process, became very significant in European iron-making in the eighteenth century especially in enabling the shift from smelting ore with charcoal to coal. Basically, it prevented the iron's contamination by the coal's sulfur.

29 It is worth noting that in 1637 Descartes published his *Dioptric, Meteors and Geometry* with the philosophical introduction, the *Discours de la Méthode*, which was to become his best-known work. This book was written in the first person in French rather than Latin and was intended to be a popular work. Descartes posed doubt in the face of empirical observation and thereby elevated mind and its construction of world.

In so doing, he fostered those forms of critical reflection of the observed that came to characterize scientific thought. In this setting Biringuccio displayed a remarkably modern mindset whereby his reflections upon his own experience from travel, observation of the work of others, his own labor and experimentation were cast against the superstitions of his culture and age.

30   Biringuccio *Pirotechnia*, pp. 68–9.

31   Ibid., p. 152. The size given in *braccia* converts to 4¼ meters high and 1¼ to 1½ meters in diameter. Once the blast furnace had been created, it became possible to completely melt iron—liquid iron became known as *blast furnace metal* and in its solid form, *pig iron*. Pig iron was cast as an unrefined material (containing iron, impurities and a high carbon context) into useable forms, and as such became known as *cast iron*.

32   Ibid., pp. 64–5.

33   Ibid., p. 174. The general (Eurocentric) historical account of iron-making goes like this: "Iron was first smelted with coal in the form of coke by Abraham Darby at Coalbrookdale (now part of Ironbridge, Telford) in 1709." J. R. Harris *The British Iron Industry 1700–1850* London: Macmillan Education, 1988, p. 30.

34   Ibid., p. 371. On this comment Smith adds the following in 1942: "It is interesting to note that the recent revival of this process on a scientific basis with the production of degrees of hardness and toughness in combinations that were impossible on the basis of quenching and reheating. For once it seems that a lost art has been rediscovered and the metallurgical products of the old masters actually were sometimes superior to modern ones."

35   Agricola *De Re Metallica*, p. xxvii.

36   Ibid., p. 615.

37   Smith's introduction to *Pyrotechnia*.

38   Needham op. cit., p. 9. In fact, in the West, the technology went backward in the Dark Ages, such that Greek and Roman iron and steel-making was more advanced than medieval methods.

39   Carbon may exist free in this iron as graphite or in a combined form as ferric carbide.

40   In early iron-making this meant drawing or blowing a draught over the surface of the molten iron while stirring (or puddling) to constantly expose the impurities in the crucible to air and thus oxygen.

41   Wrought iron was produced in the West at a lower temperature than in China, and by force rather than heat—this through the use of a forge hammer (first manual, then water-driven, then powered by a steam engine). A hot lump of iron was worked and the slag, containing the impurities, was hammered out. By a process of heating and reheating the iron in the forge and by the application of a blast of air at the moment of hammering, oxidization occurred.

42 It was found that the metallurgy of cast iron could be altered by variations in blast furnace and remelting conditions, with the result that it could be made more malleable. This advance did not occur, however, until the development of the process of "malleablizing" cast iron by the French scientist Réaumur in the early eighteenth century whereby carbon was oxidized away. Additionally, malleablizing cast iron transformed it into a machineable material, *malleable cast iron*. Contrary to some characterizations, malleable cast iron is not the same material as wrought iron—which is soft but tough, impure with very low carbon content and resulting from the oxidization of pig iron. Malleable cast iron is a material made redundant by steel, and wrought iron is still made by craft workers in very small quantities for non-structural and mostly decorative applications (like garden gates).

43 In terms of chemical composition, modern steel is divided into three groups: carbon steels, low-alloy steels and high-alloy steels. Modern steel, as an alloy of carbon and iron, can have a presence in the iron as a free element or combined in the crystalline structure of the metal. All steels contain traces of residual impurities remaining from the steel-making process. These include, dominantly, manganese, silicon, phosphorus, and sulfur. Carbon steel is the most common (representing about 90 percent of all steel produced). It ranges between high-carbon steel (above 0.6 percent carbon) to ultra-low-carbon steel of less than 0.010 percent). The other elements in these steels are of the order of 2 percent. Low-alloy steel can have up to 8 percent alloying elements—high-alloy steel is above this figure.

The elimination of impurities and reduction of carbon to a level within the parameters of carbon steel was the major driver of furnace design, research and development—to the point where, in modern, integrated steel-making, a blast furnace is charged with ore of known properties which is delivered in a variety of prepared forms as "lump" and "fines" (powdered ore), formed into "pellets" or "sinter" (powdered limestone, coke dust and fines cemented together), with a small amount of flux material (limestone, dolomite, quartzite and coke) plus preheated air—all to reduce ore to "pure iron" and load the "slag" with impurities. Thereafter the iron is able to be made into steel via a Basic Oxygen Steel-making (BOS) process that removes remaining impurities by oxidization, controls temperature with a "balanced charge" and finally adds the required alloys.

44 It should also be noted that the co-fusion method of making steel was also known to the Arabs.

45 Needham op. cit., p. 47.

46 Ibid., p. 26.

47 This steel-making technology was given its most advanced application in the first two decades of the nineteenth century by the German steel-maker Krupp, who perfected the technique in order to manufacture machine components, ship propellers and armaments.

48 This furnace combined Pierre Martin's improved regenerative principle with the furnace designed and developed by Siemens. The Siemens-Martin process functioned by hot gases being drawn through flues, which they then heated and through which the blast was channeled. This was linked to a checker work brick labyrinth. One chamber provided a passage of hot air combined with unburnt gases which increased furnace efficiency and produced a high temperature flame directed onto the charge. The other chamber exhausted burnt gases. In this furnace, a charge of pig iron, limestone and steel scrap are dumped in the hearth and heated until they fuse. Then, a large charge of molten pig iron (blast furnace metal) is added, with other fluxing material deposited later. The process could take moderately large quantities of materials and was able to be well controlled, but it was slow and expensive to operate.

49 The arrival of modern steel-making can be seen statistically—for example, in 1720 there were of the order of 300 blast furnaces in Britain (which was to become a major world iron-making nation) producing about 17,500 tons of iron; by 1820 this had risen to 400,000 tons and 50 years later (when the open-hearth was just starting to come into its own) this increased to nearly 6,000,000 tons.

50 It is perhaps of interest to contrast the difference and similarities between the writings of Chhü Ta-Chün on the description of a blast furnace (writing in 1690) with Biringuccio. First the shape and size are detailed (it is the shape of a "jar" and of the order of 3 meters high, 9 meters long and 3 meters across). It is constructed from these materials: lime, sand, salt, clay, vinegar, thick vine cables, "iron strength wood and purple-thorn wood." Its tuyères are made from "water stone" (so named because it does not burn). Chhü then says: "… it is (often) built leaning against a cliff side for greater stability. Behind the furnace is a mouth, and outside this there is an earthen wall, in which are contrived two door-shaped (i.e. hinged) bellows 5 or 6ft high and 4ft wide. Each is worked by four men, one closing while the other is opening, so as to drive a great blast (into the furnace)." Needham op. cit., p. 17.

51 It should be noted that that ore and most of the impure iron requires more heat to melt than "pure iron."

52 Ibid., p. 13.

53 On Blowing Engines see Joseph Needham *Science and Civilisation in China Volume 4: Physics and Physical Technology* Cambridge: Cambridge University Press, 1965, p. 369.

54 Needham *The Development of Iron and Steel Technology in China*, pp. 18–19.

55 Ibid., p. 23.

56 Ibid., p. 20.

57 For a brief overview of these developments see Robert Temple *The Genius of China* New York: Simon and Schuster, 1986.

58 Scrivenor op. cit., pp. 158–9.

59 Joseph Needham *The Development of Iron and Steel Technology in China: Second Biennial Dickinson Memorial* Lecture London: Newcomen Society, 1956. Needham, cites the famed ironmaster Cho Shi, who founded an iron works in Szechuan in the third century, which had a highly organized system of production and employed nearly 2,000 men.

60 For example, the *Thien Kung Khai Wu* (*The Exploitation of the Works of Nature*) of 1637, which addressed agriculture and industry and is described as "China's greatest technological classic." This material is itself linked to a series of important primary texts, like the *Khao Kung Chi* (Artificers Record), which, in turn, contained a chapter of the *Chou Li* (Record of the Institutions of the Chou Dynasty)—the original of this document was lost at the beginning of the Han Dynasty and a substitute document was collected by Prince Hsien of Ho-Chien in the second quarter of the second century. See Joseph Needham *Science and Civilisation in China Volume 4, Physics and Physical Technology* Part 2, Mechanical Engineering, Section 27, p. 18. Note all dates specified are based upon a Western Judeo-Christian calendar—which itself makes a point on the non-availability of a neutral point of reference.

61 These departments, while subject to occasional changes of name, endured over many hundreds of years.

62 For an account of this measurement see Lothar Ledderose *Ten Thousand Things: Module and Mass Production in Chinese Art* Princeton, NJ: Princeton University Press, 2000, p. 134.

63 Liang Sicheng *Ying Zao ta Shi Zhu Shi* (Volumes 1–13 of a total of 34 sections) Beijing: Zhong guo Jianzhu, Gongye Chubabshe, 1983. This facsimile edition, based on the first modern translation of 1925, was the product of many decades of research and heralded the beginning of modern Chinese architectural history. The latest edited, with a new introduction, was produced in 1963; however, it was kept hidden during the course of the Cultural Revolution and not published until the early 1980s.

FIGURE 2.1 Charcoal-makers, England c.1900. Photographer unknown.
Authors' collection

2 ECOLOGIES OF CARBON

I n our first chapter we gave a sense of the complexity of the immaterial "ecologies of mind" by which the knowledge of steel-making was elaborated as it travelled between Eastern and Western cultures over extended periods of time. In this chapter we will start to build an image of a particular material ecology by focusing on the fuels used in iron and steel-making—wood, charcoal, coke, coal and to a lesser extent, oil—and some of their environmental impacts.

Basically, fossil fuels provide both the means of generating heat and the source of carbon, the key alloy element within iron in the steel-making process.

In a period of less than 40 years, the world consumption of iron ore, coal, clay, limestone, water and many other materials of steel-making has more than doubled.[1] Notwithstanding attainments in impact reduction by a significant number of steel-makers worldwide, environmental impacts of the industry continue to grow—this because an expanding global market is propelling volume of production. At its most basic, one of the fundamental material causes of structural unsustainability is ecologically unsupportable growth. This is a problem that demands to be faced.

Again we will focus on China as a means to bring a Eurocentric perspective into question and because it also affords the opportunity to "learn from difference."

The tree of China

From ancient times in China the tree had great significance as an ecological indicator. It supplied knowledge of "the *Tao* of the earth"—in other words, its health mirrored the health of the earth itself (whether "the earth" is deemed to be local soil conditions or the entire planet).

Because of China's early development and the refinement of its administrative system, there is a substantial historical record of its practices of environmental administration, not least by what we now call "natural resource management." The *Chou Li* provides one of the earliest records of the environmental management of imperial domains in ancient China and identifies the roles of various functionaries. Over 2,000 years ago, there were Superintendents of the Mountains and Marshes, Inspectors of Forests and Rivers and a Controller of Charcoal. The duties of the Inspector of Forests included designating which tree species in the forest could be cut and when. He also kept a tally of animals in the imperial herds and recorded the numbers killed.[2] Areas of responsibility for rivers and forests sometimes overlapped. For example, willows and poplars had been planted as long ago as the −sixth century to assist water management and dike stabilization.[3] The environmental management of forests became increasingly sophisticated and continued to be developed right up to the eleventh century.

Forestry in China has an ancient history, and references to clearing forests for agriculture go back many thousands of years. Deforestation was already a problem in northern China 3,000 or even 4,000 years ago. Bronze and iron smelting along with unrestrained harvesting were responsible for worsening this condition. As result of this created scarcity, timber had to be transported vast distances from the south to the north—a situation that is still the case today.[4] Pine plantations were established in the eleventh century to redress the problem of a shortage of construction timber. These trees were frequently planted in areas where other species of trees would not grow (like high in the mountains).[5] Hardwoods were also planted over an extensive period of time—for weapons, building materials, civil construction and charcoal production. Elms were planted in border regions as part of the nation's system of defense, as well as for firewood.

By the −first century the state was taking action to ensure cities had sufficient supplies of wood, while forest management had become an established practice with its own body of detailed knowledge of tree species, their uses and cultivation. Paulownia was cultivated to stabilize soil after landslides, as an edging after the cutting of roads and as a crop on abandoned agricultural land; the light, insect-resistant and durable wood was used for making hives. During the eleventh century Chên Chu wrote a ten-volume work of instruction—a text as comprehensive as any modern monograph on a forest tree species.[6]

Besides plantation establishment and management, the most advanced and celebrated Chinese forest practices were coppicing and pollarding. Oaks were a favored tree and were coppiced and pollarded for charcoal manufacture and fuel wood. Coppicing is a practice based on cutting to the stump from specially planted stands of trees, then allowing regrowth. A crop was taken on an eight-year rotation, a process that continued over several generations. Thinnings were taken for fuel, selected shoots were allowed to regrow and side growth was regularly trimmed during culti-vation and prior to harvesting. With pollarding, the stump is allowed to regrow to about two meters prior to major member selection—it delivers an intermediate crop while waiting for the main crop to grow to cutting size. Additionally oak leaves, along with those of the mulberry, were used as feed for silkworms.[7]

As indicated in Chapter 1, while coal was substituted for charcoal in the smelting of metals in China very early in comparison with Europe, this did not become the major practice until the eleventh century. Subject to local availability, a considerable amount of charcoal was used. However, the ratio of supply, demand and output suggests that charcoal production never became a cause of deforestation on the same scale as it did in Europe. Forest administration was in fact mainly driven by agricultural land clearing and the need for construction timber. Menzies' review of the literature on the topic is inconclusive.[8] It is also important to recognize the vast span of time covered by these events and their episodic nature. We are looking at activity over at least a 2,000-year period, and events certainly do not follow a single developmental direction. The ebbs and flows of economic and political circumstances in China, as well as significant regional differences, meant that a true picture only arrived as a mosaic. Thus, our view can be no more than very impressionistic.[9] However, what is not in question, as Menzies clearly shows, citing among others the great fourth-century philosopher Mencius lamenting the transformation of Ox Mountain by deforestation, is that by the opening of the first millennium deforestation was already a substantial problem in all Chinese dynasties. By the twelfth century, the patterns that are still evident today were firmly established—like flooding due to siltation upstream from the Yangtze estuary at times of major water discharge.[10] As we have seen, at the same time of the lament by Mencius, and perhaps long before anywhere else, ancient Chinese culture was recognizing the need to manage environmental impacts.

Roots of the European problem

The European story of forest ecologies and iron and steel-making is starker than that of China.[11]

Iron-making in Europe was, for hundreds of years, a forest industry that was almost completely dependent on charcoal. The forest location was vital because charcoal was a fragile and friable material—transporting it over rough roads in carts rendered it to dust. Even in the well-organized system in Sweden, which for a long time was Europe's largest and highest quality iron-maker, the problem was not surmountable. The Swedish method was to fell trees in the spring, let the wood dry over summer and then make charcoal in winter, so it could be transported over snow by smooth-running sledges. Handled carefully, the maximum distance charcoal could travel was about 24 kilometers.[12] This prompted a system whereby furnaces were torn down and newly built ever deeper into forests as trees were felled. The landscapes and biophysical ecologies of Austria, Britain, Belgium, France, Germany, Italy, Norway, Prussia, Russia, Spain and Sweden were all by degree altered by this activity.

These methods were maintained over an extensive period of time. One of the clearest examples can be seen in the Forest of Dean, in Gloucestershire, England. This was a major location of iron ore mining and smelting, first established by the Romans, then re-established in medieval times as an important center of iron production. By the turn of the tenth century, the nearby city of Gloucester had become a center of forging and trading in iron.[13] Scrivenor notes that an iron and wire works established near Tintern Abbey in the Forest of Dean in the sixteenth century was the work of Germans (as creators of some of the first European blast furnaces, the Germans also became exporters of skills).[14] In contrast to charcoal, iron itself was traded over considerable distances—for example, by the eleventh century Germany was exporting iron to many European nations, including Britain.

Deforestation became a significant problem in Europe from the fifteenth to the end of the eighteenth centuries, although the problem had been recognized long before then. Plato, for instance, wrote in graphic detail in *Critias* of the destruction of soil fertility in the hills, high crests and rocky plains of Phelleus caused by the clearing of thick woods from the mountains. This occurred at the time of prehistoric Athens, creating a situation in which "… the rich, soft soil has all run away leaving nothing of the land but skin and bone," and where once "… the soil benefited from an annual rainfall which did not run to waste off the bare earth as it does today …"[15]

European deforestation was not only the product of the iron industry's demand for charcoal but also of the fuel requirements of glass and brick makers. Additionally, shipbuilders' needs for timber, especially oak, had major impacts on forests. The deforestation problem was sought to be dealt with by legal means, by estate management and through the creation and codification of a body of knowledge on silviculture.

From the fifteenth to the end of the eighteenth centuries, Britain led forest destruction in Europe. This was due to a convergence of five factors:[16]

- proportionally its population was growing faster than any other European nation (while general population doubled in this period, the populations of cities increased eightfold);

- as a small island its forest resources fell short of demands for timber;

- it created a large navy of timber ships armed with iron cannon;

- its iron-making was expanding (not least because of the need for cannon); and,

- its development of industry was the fastest of any nation.

Such was the extent of deforestation in England that by the mid-sixteenth century Acts of Parliament were passed attempting to curb the level of destruction.

In 1543 Henry VIII, whose creation of the Royal Navy significantly increased the number of warships being built, started regulating the cutting of coppices for charcoal-making in the counties of Surrey, Sussex and Kent. A Parliamentary act was passed in 1558 (the year Elizabeth I came to the throne) directing that timber should not be felled "to make coals for burning iron." It stated that "no timber, of breadth of one foot square at the stub, and growing within fourteen miles of the sea, or of any part of the rivers Thames, Severn, or any river, creek, or stream, by the which carriage is commonly used by boat, or other vessel, to any part of the sea, shall be converted to coal, or fuel, for the making of iron."[17] Other statutes (which were all ineffective due to poor enforcement) followed in 1581 and 1585. By 1630, deforestation had become a problem not just in West, Weald and Southern counties but had spread as far north as Durham.[18] Over the next 40 years it is claimed that ironmasters were to become dependent upon charcoal produced from coppicing; even so, this practice was unable to keep pace with the demand for timber, with the result that iron production dipped between the end of the age of charcoal and the establishment of coke as the replacement fuel and carbon source.[19]

What needs to be grasped is just how much timber charcoal-making consumed. The calculation is notoriously vague, as so much depended on the quality of the wood, the method employed, the skill of the ironmaster as well as the design, size and efficiency of the blast furnace. Thus inefficient iron-making at the start of the seventeenth century required more charcoal to make the same amount of iron than by the mid-eighteenth. With these qualifications, we note that a Swedish study estimated that by the end of the late eighteenth century 1 ton of finished bar iron required over 400 hectoliters (4,000 liters) of charcoal (a century earlier it would have require another 100 hectoliters), which translates to several tons.[20] As anyone who has handled a lump of charcoal knows, it is extremely light, because all fluids are extracted in the course of its manufacture. The same study estimated that 1 ton of charcoal represented the continuous labor of one worker for two and a half months.[21]

The general view was that British Parliamentary legislation failed to resolve the problem—and it was not resolved until coal was turned to coke and used to smelt iron in the seventeenth century.[22] Interestingly, Aitchison cites a letter written at the time by Abraham Darby (the first European ironmaster to perfect iron-making with coal in 1708/9):

Had not these discoveries been made the Iron Trade of our own produce would have dwindled away, for woods for charcoal became very scarce and landed gentlemen rose the price of cord wood exceeding high—indeed it would not be got.[23]

It is in this context that we can view the 1719 Bill on the importation of timber from the "British American Plantations" which specified that ships returning to British waters that were not fully laden with goods like sugar and tobacco had to carry timber. Besides supplementing local timber supply, the Bill aimed to restrict industrial development in America. It went so far as to prohibit the importing of iron from the colony. The enacted Bill stayed in place until 1750; however, its result was to reduce, rather than direct the nature of, trade.[24] Once the American steel industry gained momentum in its own right, an enormous wave of deforestation commenced, resulting in the destruction of hundreds of thousands of hectares of forest.[25] Closer to home, and more dramatically, the shortage and cost of timber in England was responsible for the mass destruction of Ireland's forests during a 30-year period (1672–1703). The taking of timber from Ireland was undertaken not only for profit and fuel, but also to deprive "the banditti of their lurking places," thus deforestation became a military expedient.[26]

Such was the situation in Ireland that during the reign of William III an Act was passed requiring a quarter of a million tress to be planted in Ireland; moreover, owners of iron works were required to plant 500 trees annually.[27] However, poor economic and political circumstances meant that stocks of charcoal continued to diminish and Irish iron-making ceased.

How do we view environmental impacts during this period?

Clearly, the destruction of old growth forests and their biodiversity, as well as totally uncontrolled emissions from charcoal and iron manufacture, were significant, especially as the Industrial Revolution, which was a revolution of iron, gathered pace. Besides being a key moment in escalating human-induced climate change, there were significant transformations of landforms, land drainage, soil nutrient levels, wildlife populations and plant species. The consequences of these changes are still evident in many parts of Europe, most noticeably in the now uncultivable and almost inert moorland soils of England, Scotland and northern Germany.

By degree, the removal of trees in quantity diminishes the total volume of oxygen produced and reduces atmospheric moisture. There are also related impacts which can include:

(i) the de-sequestering of soil-bound carbon;

(ii) reduced soil fertility by the elimination of nutrient sources;

(iii) initiating processes of continuous soil erosion (top soil is washed away, nutrients are leached out, rivers become silted and the incidence of flooding increases);

(iv) a rise in the water table by a reduction of deep root water take-up;

(v) reduced ground water movement and drainage into water courses (this and the prior factor can combine in various ways resulting in land that is permanently waterlogged);

(vi) rising ground water prompting ground-held salts to rise (soils become salinated and water turns brackish) and in some circumstances, exposure and activation of soil-bound acid sulfates (rendering soil sterile from high acidity and resulting in land impossible to utilize except for grazing sheep in dispersed flocks); and finally,

(viii) reduction of local biodiversity, which can have a profound impact on the food chain and species viability.

While such detail, and earlier comment on deforestation, is the stuff of historical ecology, and an area we will only but touch on, it is clear that ecological problems, and related unsustainable practices, were well entrenched a very long time before the European idea of ecological systems was coined by the German biologist/zoologist Ernst Haeckel in 1873. There was then a further time gap between the arrival of the idea of the ecological and awareness of how human actions on a large scale have caused, and continue to cause, ecological damage.

The rise of harmful industrial practices not only changed the material and social environment, they also transformed physiologies, values, the way people thought about historical time and the future.[28]

Fuels: Histories and futures

To be able to make steel while also significantly reducing greenhouse gas emissions is an enormous challenge for the industry. Types and uses of fuel are crucial here. There are four determinate factors that articulate uses of the fuels in iron and steel-making: cost; availability; carbon content; and use without the contamination of the metal. As indicated, coke, coal and charcoal are not simply used as energy sources to create the levels of heat needed to smelt iron and make steel, they are also essential to thermochemical processes (not least in the production of gases required for process operation, the extraction of oxygen from ore and the control of carbon content).

Wood and charcoal

Wood, the first fuel, still remains in use. Initially, and over an extensive period of time, European iron-making utilized dried wood as a fuel (along with charcoal) not least because it was cheap. However, it is not an efficient fuel. The volume-to-weight ratio for produced heat is very poor. Nevertheless, today in some countries, because forest waste and to a lesser extent wood chips are cheap, wood is still used (it is also being used as a fuel for electricity generation).[29]

Large-scale supply of wood chips as a fuel stock for fuel-intensive industries, like energy generation and steel-making, would however have massive negative environmental impacts, not least in terms of greenhouse gas emissions and deforestation. Likewise, using high-grade plantation-produced timber for fuel stock is a completely unsustainable practice. Renewable annual biomass crops like straw are also talked of as potential fuel stocks, but they present contradictory options. While not having as high an impact as timber, one questions if these materials could be supplied in sufficient quantities. Of course, they equally pose a problem of CO_2 emissions.

As Biringuccio tells us, charcoal was the primary fuel of early Western iron-making.[30] Charcoal-making became an established industry in Europe over 3,000 years ago. It was the primary fuel in the smelting of iron ore, and the ores of other metals—notably tin, silver, gold and copper. As we will show, it is not only part of the past of iron and steel-making but also part of its future.

Charcoal was, and is, a far more efficient means of the utilization of wood as a fuel. The carbonization process by which it is now made enables byproduct extraction and the carbon ratio of the material to be raised.[31] Carbonization processes have developed considerably since the initial manufacture of charcoal in pits and earth mound kilns in the distant past. Production is now faster and far more efficient. It now takes place in closed system retorts where material quality, off-gassing and the extraction of pyrolysis oil are all completely controlled.[32] The byproducts of charcoal-making, although now more refined, have always been significant. In the pre-industrial and early industrial history of European iron-making, the period when deforestation was rampant and dominated by the demands of charcoal-making and shipbuilding, tar extracted during carbonization was sold on to the shipbuilders for caulking (sealing seams in planking).

By the seventeenth century, chemists had discovered that the condensates of charcoal, the pyrolysis oils, were a valuable source of raw materials, the first being acetic acid discovered by Rudolf Glauber in 1635.[33] Others included organic acids, methanol, aldehydes, acetone and creosotes. As kilns and the quality of charcoal they made improved, mostly as a result of German and Swedish innovation, the ability to efficiently extract condensates increased. So while the significance of charcoal as a fuel stock diminished as it was displaced by coal, the industry did not totally die. In fact, it became a significant raw material for the manufacture of plastics, was used in various chemical processes, especially in the pharmaceuticals industry, and was utilized in a variety of filter industry technologies (the best-known of these being in water filtration and gas masks).

It is worth spending a little more time looking a little more technically at some at the characteristics of modern charcoal blast furnaces and their potential.

The contemporary uses of charcoal no longer depend on timber of a standard size. Briquetting, pelleting and blowing charcoal dust into a furnace are all current methods that increase flexibility in the use of the material.[34] Historically, one of charcoal's major limitations has been its poor weight-bearing ability, which is extremely significant in determining the size of the charge of a blast furnace, and thus the size of the furnace itself. A furnace charged with too much weight of ore would crush the charcoal, restrict airflow and dramatically reduce the furnace operating temperatures. This means limiting the total capacity of the furnace to about 1,200 tonnes, which is six to ten times smaller than most modern coke blast furnaces. The structural properties of the timber from which charcoal is made vary considerably—in the past, oak, beech, ash, chestnut and (in the USA) red maple were favored, although many other types of trees were used. Today, the strongest and most favored timbers, in terms of their moisture content and volatiles, are eucalyptus species (all of which are seeded from Australian stock). Briquettes are another fuel option, as they have the ability to increase the structural strength of charcoal; however, they come with the disadvantages of greater cost, increased emissions from the combustion of the binding material (e.g. molasses) and an increased volume of slag.

The charcoal blast furnace functions at a lower reserve zone temperature than a coke blast furnace.[35] This means that the furnace refractory lining is not subjected to as much punishment and thus does not need to be of the same specification. Equally, the charcoal blast furnace operates with just shell cooling, not requiring a supplementary cooling technology. Additionally, a lower blast temperature is required which means that blast pre-heating can be delivered with tubular heat exchangers rather than with stoves, and blast pressure can be provided with centrifugal blowers rather than with very costly turbo-blowers.

Because charcoal is about half the density of coke, the ratio of ore to charcoal is about 1:6 as opposed to 1:3 for ore to coke. There are a few other linked factors: a charcoal blast furnace does not require sinter, can run on lump ore with a considerable variation in lump size; however it does require an ore which is highly reducible. At the same time, charcoal produces less ash than coke, and thus creates a low ash slag that is about half of the quantity created by a coke blast furnace.

All of these factors combine to make a charcoal blast furnace comparatively cheaper.

Charcoal in the future of steel-making

Charcoal has the potential to become the basis for significant forest-based economic activity as an alternative to destructive practices such as wood-chipping of old growth forests; it could also make a substantial contribution to reducing the steel industry's environmental impacts beyond crucial CO_2 emissions reduction. As a renewable resource, charcoal has the potential to contribute to biophysical sustainment and added environmental value.[36]

Charcoal manufacture demands a relatively high labor-intensive workforce, which means it has a significant employment multiplier effect. Besides reversing the decline in the economies of forest communities, an appropriately designed and managed charcoal industry has the ability to drive reforestation, advance sustainable forest management and produce a fuel stock with much lower environmental impact than coke—providing the production of charcoal is carried out in advanced retorts. As both historical and contemporary evidence indicates, high-quality iron can be made with charcoal. Sweden, which ran charcoal blast furnaces up until 1966, produced the highest quality iron in Europe for several hundred years. The Swedish Royal Institute of Technology conducted research for an extensive period of time (1927–64) on the chemistry of charcoal blast furnaces.[37] Brazil, currently the world's leading charcoal blast furnace steel-maker, produces quality product. The use of charcoal is now being claimed as making a significant contribution to reducing the environmental impact of steel-making—especially in Brazil.[38] Argentina also has a modern history of charcoal blast furnace operation, as does India and Australia (in the latter, research into advanced practices is being carried out by industry in cooperation with a State Department of Forests).[39]

The most efficient and environmentally responsible charcoal production technology is the modern "Batch or Continuous Retort." While these retorts vary in cost, structural design, capacity, operation methods, labor and maintenance requirements, they all produce charcoal of a high quality (which means density, structural strength and a fixed carbon content of the order of 90 to 95 percent) in fully controlled conditions that use the combustible gases and recover volatile content.

Probably the most appropriate and progressive use of charcoal in modern steel-making is as the fuel and carbon source to power a mini-blast furnace able to be used in conjunction with an electric arc furnace (EAF). The EAF is the core technology of the mini-mill, and (as we will see in more detail later) has been the fastest growing method of steel-making in recent times, having brought many "developing" countries into the steel supplier community—this because the cost of purchasing and operating a mini-mill is a fraction of the enormous cost of an integrated steel works. EAFs are predominantly scrap-fed, and thus vulnerable to fluctuations in availability and cost of scrap steel. What a small and reasonably cheap charcoal blast furnace can do is to bring iron-making within the economic and practical reach of mini-mills that lack a continuous supply of scrap steel. This facility does not necessarily displace scrap utilization but ensures a continuity of iron feedstock for the EAF when scrap supply is intermittent or its cost very high.

FIGURE 2.2 Coke ovens built in 1906, Shoaf Mine & Coke Works, Fayette County, Pennsylvania. Courtesy of US Library of Congress, Prints & Photographs Division. Reproduction number: HAER PA, 26-SHO, 1–3

Clearly, a significant issue for charcoal-based steel-making is how the trees for charcoal can be grown and harvested sustainably. They should not displace productive agricultural land; neither should tree harvesting be done in such a way as to reduce land fertility. An appropriate land use strategy would underake reforestation of degraded land to redress soil erosion, improve levels of soil nutrients, support increased biodiversity and contribute to carbon sequestration. This requires research, policy development and commitment. Certainly, such research would need to identify low-impact methods of: planting; plantation management (including weed/pest control); harvesting; soil nutrient and CO_2 retention; wildlife protection or relocation; as well as innovative approaches such as the use of organic waste like sludge as a soil conditioner or growing medium.

Coal and coke

We noted that coal was employed early in the history of iron-making in China and late in the history of the West. In the early Chinese crucible smelters, coal was used as fuel but not in the reduction process. For this to happen coal has to be turned to coke, which removes the sulfur, the major contaminant.

There are many types of coal, each with different properties and ratios of materials especially ash, volatile matter, sulfur (coal contains sulfur up to the order of 2.5 percent) and fixed carbon. Iron-making favors coke made from bituminous coal, with a 70 to 90 percent carbon content or the less available anthracite with 95 percent carbon content.

The environmental impact of coal utilization was less immediately evident than the deforestation associated with charcoal production. However, while the respective impacts on the ground are visually very different, both have contributed to climate change. If iron and then steel were the primary materials of the first machine age (the Industrial Revolution), coal was its fuel.

Of course, it is difficult to isolate the environmental impact of burning coal in blast furnaces from its use by other industries, for domestic heating, and, historically, as a transport fuel for shipping and rail. The most immediate impact, especially over the course of the nineteenth century, was on the air quality of industrial cities. Life-threatening smogs were common in advanced economies right up to the mid-twentieth century. The health of populations living in the industrial areas of many "developing" countries is still damaged by direct contact with this kind of pollution.

What took much longer to be recognized by the global scientific community, and people at large, was that burning fossil fuels creates emissions of gases, especially CO_2, that accumulate in the outer reaches of the planet's atmosphere, trapping rather than dissipating reflected heat. Here then is the greenhouse effect, and its contribution to global warming. While in the short view, the temperature increase is not dramatic (i.e. two to four degrees over a century), the consequences will be profound—and are still only partly grasped. Rising sea levels, increasing ocean temperatures, changes to the volume of fresh water held in the polar ice caps, alterations to rainfall (drier in some regions, wetter in others) leading to shifts in farming and harvesting patterns, habitat changes from bacterial to animal levels—these are just some of the factors that are still being attempted to be fully understood. Undoubtedly, there are other consequences yet to emerge. It is true that there have been other events in the planet's distant past, like massive volcanic eruptions, that have had major consequences for the climate and biological life. But what such events confirm is that while the planet has withstood massive geological trauma, organic life is fragile. There is now, however, no argument about the significance of the impact of human-led actions. Humanity urgently needs to address the problems it has created. That all "the facts" are not fully known is no reason not to take action, as the global program of emissions controls, even in its insufficiency, affirms.

The "breakthrough" that enabled coal to be turned to coke is iconically associated with Abraham Darby's efforts of 1708/9 (the actual date is contested). Darby's attainment was technology transfer rather than an invention. Most likely what he did was to transfer to iron-making the practice of English brewers, who used coke for drying malt. Like charcoal, coal is subjected to a carbonization process, and coke is the result. This is carried out in coke ovens, which became an essential part of an integrated iron and steel works.

Making coke has many similarities to charcoal-making—volatiles (mainly light oil, sulfur, tar, phenol and ammonia) are extracted and the remaining pasty material forms into porous lumps of pure carbon. The key requirements of coal for coking are that it should swell, become plastic, solidify and subsequently shrink during further heating—performance in these areas defines the suitability of particular kinds of coal for coke-making.

Coke was initially made in beehive ovens that simply discharged steam and smoke charged with sulfur fumes, particulates, plus a whole cocktail of chemical emissions over a period of about six days. The capture of volatiles as byproducts did not become widespread until the very end of the nineteenth century, and while this reduced emissions, there was still enormous air pollution, water pollution and soil contamination (the latter because of acid rain from sulfur dioxide emissions). Water cooling (by quenching) also added to air and water-borne emissions. Modern coke-making now takes less than one day in ovens that function with significantly reduced emission levels. The most advanced retorts are of the same order of performance as those making charcoal.

Variations in the chemical composition of coke directly influence the reduction of iron in a blast furnace, the elemental composition of the metal and the qualities of blast furnace slag.[40] Coke is tested for its purity, with particular emphasis on its major elements: carbon, hydrogen, the percentages of nitrogen and organic sulfur directly linked to another key element—oxygen. The relevant ratio of the major elements directly influences its calorific value. Other elements found in the mineral matter of coal and coke can include chlorine, phosphorus, arsenic and alkali metal. These elements determine the nature of the iron and steel produced (along with other process materials, e.g. flux materials) and the refractory materials of the furnace.

Coke is thus crucial to the thermochemical interactions that occur between iron ore, carbon and oxygen in a blast furnace. Importantly, the ability of coke to retain its "structural strength," remaining solid during its passage through the melting zone of the blast furnace and its porosity which assists the air flow of the blast, are both features that have secured its dominance in the industry over an extended period of time from the steam-powered blowing engines of the first industrial age to present-day steam turbines driving blast furnace turbo-blowers.

In contemporary integrated steel-making, coke production and use are subject to two developments. First, in responsible nations, coke-making is being made to conform to more stringent environmental regulation, one of the major improvements having been the reduction of leakage from oven doors; another is the introduction of dry quenching which yields less pollutants than wet quenching. Second, the volume of coke used is able to be reduced by the use of pulverized coal injection (PCI) or the injection of natural gas (NG). These technologies are features of modern blast furnaces and are also employed in retrofitting existing blast furnaces that have not had a prior conversion to oil injection.[41]

Notes

1   See Gary Gardner and Payal Sampat *Mind over Matter: Recasting the Role of Materials in Our Lives* Danvers, MA: Worldwatch Institute Paper 144, December 1998, p. 16.

2   Joseph Needham *Science and Civilisation in China Volume 6*, Part 3, "Forestry" (by Nicholas K. Menzies) Cambridge: Cambridge University Press, 1996, p. 611.

3   Ibid., p. 591.

4   Ibid., pp. 561–2.

5   Ibid., p. 574.

6   Ibid., pp. 596–7.

7   Ibid., pp. 601–2, 621–2.

8 Ibid.

9 Iron and steel-making in China has ranged from small-scale and scattered activity, to small-scale activity organized into large clusters or even systems, to large-scale, big workforce consolidated operations. Metals have been smelted in plants employing thousands of workers and equally in back yards with small furnaces with just a handful of workers: such practices were occurring both a few thousand years and a few decades ago. It is against this backdrop that we can view, for example, large-scale logging in the border regions of Shensi, Ssuchhuan and Pei in the eighteenth century employing tens of thousands of men and undertaken to supply fuel for iron foundries in the mountains and timber for cities in the lowlands. Ibid., p. 562.

10 Ibid., pp. 650–4.

11 On the procurement of wood and charcoal Vannoccio Biringuccio writes "… very great forests are found which make one think that the ages of man would never consume them in such uses, especially since Nature, so very liberal, produces new ones every day. But what need is there to speak of quantity? Are not the mountains covered with trees, the valleys full, and the plains occupied? Do not wild trees outnumber even the leaves of domesticated ones, and the areas occupied by trees exceed perhaps those that are free? Certainly I believe that men [will be unable to smelt] more as result of a lack of ores than because they are unable to use fire on account of the great consumption of such material." Vannoccio Biringuccio *Pirotechnia* Cambridge, MA: The MIT Press, 1966 (reissue of the 1942 edition), p. 174.

12 This information is drawn from published material supplied by the Swedish steel organization, Jernkontoret under the title "Swedish iron in the seventeenth and eighteenth centuries. Export industry before the industrialisation." We also note that America in the late nineteenth century "solved" the problem of transportation by tightly packing the material in specially designed rail cars—this worked well, with the exception of occasional fires ignited by charcoal that were not fully extinguished.

13 Harry Scrivenor *History of the Iron Trade: From the Earliest Records to the Present* London: Longman, Brown, Green and Longmans, 1854, p. 32.

14 Ibid., note p. 35.

15 See Plato *Timaeus and Critias* (trans. Desmond Lee) Harmondsworth: Penguin Books, 1977, p. 134.

16 Robert Raymond *Out of the Fiery Furnace: The Impact of Metals on the History of Mankind* Melbourne: Macmillan, 1984, p. 148. It should be noted that a large timber warship consumed a vast amount of trees. To take one iconic example: it took 2,100 tonnes of oak to build Nelson's flagship HMS *Victory*—requiring the felling of an entire oak forest. Ibid., p. 149.

17 Scrivenor op. cit., p. 35.

18 Howard G. Roepke *Movements of the British Iron and Steel Industry—1720–1951* Urbana, IL: University of Illinois Press, 1956, p. 4. This picture of events has been subject to valid criticism in that it fails to acknowledge the complexity of a situation wherein labor costs, skills, available water power, imports and transport costs were all factors. Notwithstanding this, and while economic impacts are contestable, the environmental picture remains intact and grim.

19 Ibid., p. 5.

20 Jernkontoret "Swedish iron in the seventeenth and eighteenth centuries. Export industry before the industrialisation," pp. 85–6. We also note that the Swedish College of Mines issued licenses on the basis of quotas, which while ensuring owners of iron works having rights to forests for charcoal production, did so in a highly regulated regime. Scrivenor op. cit., pp. 151–3.

21 Around that same period in England Scrivenor tells us that there were 800 furnaces, forges and iron mills in England. The furnaces were producing in the order of 15 tons of pig iron each, with the forges making 3 tons of bar iron (which required three tons of charcoal). Ibid., p. 55.

22 In contrast the French, in the seventeenth century, used optional and financial means of control rather than political instruments. Ibid., pp. 177–8.

23 Leslie Aitchison *A History of Metals* (Volumes 1 and 2) London: MacDonald and Evans, 1960 Volume 1, p. 443.

24 Scrivenor op. cit., pp. 71–2.

25 Robert B. Gordon *American Iron 1607–1900* Baltimore, MD: Johns Hopkins University Press, 1966, pp. 40–9.

26 Ibid., p. 61.

27 Ibid., pp. 67–8.

28 See Tony Fry *A New Design Philosophy: An Introduction to Defuturing* Sydney: UNSW Press, 1999.

29 Forest waste is of the order of 20 times cheaper than charcoal, while wood chips are about 12 times cheaper. Wood does not, however, have the ability to reach the temperatures of charcoal, coal or coke.

30 Biringuccio in his *Pirotechnia* reviewed "Varieties of Charcoal" and its manner of manufacture, pp. 173–9.

31 The carbonization process is one of dry distillation where matter is raised to a high temperature and oxygen is excluded or air intake is very restricted. There are three temperature regimes: 100°C to 170°C, in which loosely bound water in the wood is evaporated; 170°C to 270°C, in which off-gassing of carbon monoxide (CO) and carbon dioxide (CO_2) and condensable vapors occurs, which after scrubbing and chilling, form pyrolysis oil; and 270°C to 280°C, in which CO and CO_2 emissions cease and condensable vapors increase—temperature increases in this phase by a spontaneous exothermic reaction. Walter Emrich *Handbook of Making Charcoal: Traditional and Industrial Methods* Dordrecht: D. Reidel Publishing Company, 1985, p. 5.

32 See Luis A. H. Nogueira, "Criteria and Indicators for Sustainable Woodfuels" in J. Domac and M. Trossero (eds) *Industrial Charcoal Production* Zagreb: Institute Acende Brazil/FAO, 2008.

33 Emrich *Handbook of Making Charcoal*, pp. 1–2.

34 J. Domac and M. Trossero (eds) *Industrial Charcoal Production*, p. 7.

35 The charcoal temperature is 800°C rather than 950°C for coke.

36 Because of the extraction processes and technologies inherent in its modern manufacture, charcoal, in comparison with coke, significantly cuts CO_2 emissions. How this reduction is calculated in a life-cycle picture is, however, complex and contestable—for instance, carbon sequestered in plantation growth can be factored in, but at present there is no consensus among scientists on the measurement of amounts of carbon actually sequestered as planting disturbs the soil and releases CO_2 so this has to be offset against the amount sequestered before any credits can be carried forward. There are claims of a CO_2 reduction of charcoal over coke to the order of 120 percent per metric tonne (how this has been calculated is, however, not clear, especially in terms of land use and soil-held carbon). Even if this figure were optimistic, it is clear that there is a very significant reduction.

37 Cited in detail by Gordon *American Iron 1607–1900*, pp. 116–18.

38 In April 2003 the largest steel-maker in the USA signed an agreement with the Companhia Vale do Rio Doce to construct and operate a pig iron production facility in northern Brazil based on two charcoal-powered mini-blast furnaces. Charcoal will be produced from local eucalyptus plantations, which are being extended from 33,200 to 81,000 hectares (*Steel News* April 25, 2003). Prior to this, it was announced that V&M do Brazil, a subsidiary of the steel maker Vallourec, is cultivating 140,000 hectares of eucalyptus plantations in the northern Minas Gerais region of Brazil for charcoal—this, in compliance with Forest Stewardship guidelines and ISO 14000 Environmental Management Standards. The CO_2 reduction claim of the use of charcoal is 1.8 per 1 tonne of iron. V&M do Brazil are refurbishing their 1600 charcoal furnaces to reduce emissions (*Steel News* February 22, 2003).

39 Ibid., pp. 71–3.

40 We also note that heat recovered from coke ovens, blast furnaces, sinter plants and basic oxygen steel-making (BOS) furnaces is able to be used to co-generate electricity. Equally, gas turbines that generate electricity can be powered from off-gases from the steel-making process.

41 Oil was also injected into now outdated open-hearth furnaces.

FIGURE 3.1 Wellcome Library, London, Medieval Alchemist at work, Diorama. By Ashenden, after E. L. G. Isabey. Collection: Wellcome Images. Library reference no.: Slide number 1101 (M0005410). Copyrighted work available under Creative Commons Attribution only licence CC BY 4.0

3 MAGIC, ALCHEMY AND SCIENCE

H aving introduced the general idea of "ecology of mind" in Chapter 1, we now consider this idea specifically in the context of an "ecology of steel."

The magic of metal

Metallurgy, the applied science of metals, travels back into the history and prehistory of science. Historically, knowledge of the gods, of the mystical and of sorcery was interwoven with the development of knowledge of materiality and ideas of scientific causality. In fact knowledge of metals was first constituted in narrative mythology and magic, then later advanced through the theory and practice of alchemy—a discourse which travelled in parallel with science for much longer than is generally acknowledged.

Although modern science views itself as vastly superior to magic, it actually stands, as we shall show, on a prehistorical foundation of knowledge anchored in symbolism and mystical world views. While this moment was prior to the modern division of knowledge produced by the Enlightenment—a division that created the schism between science and philosophy—and while the pre-scientific mode of knowing was often flawed, it was also in some respects in advance of what displaced it. At the core of many pre-scientific methods of knowing was a relational, rather than teleological (directional) view of causality. This recognized that seemingly unconnected and incommensurate registers of the material and immaterial could determine each other. This, as we will demonstrate, is of great importance even though, from the perspective of modern science, many of the conclusions reached were in error. This recognition of relationality runs directly counter to the modern view of specialization as the path to true knowledge (expressed as "the ability of modern scientists to know more and more about less and less"). So much modern thought fails to see the relational impacts of scientifically directed actions. Yet there remain some continuities between ancient and modern thought, like the quest for magical materials with supra-qualities and the desire to create wealth beyond measure.

One of the continuities shared by magic, alchemy and science is the power of secret knowledge communicated by esoteric and exoteric symbolism.[1] In the distant past, symbolic forms were used with allegorical references to mythologies. These forms become iconic (abstract to all but the few) and united the mystical with the material. They expressed the power of the owners of the concealed knowledge to maintain the general condition of unknowing; they equally concealed the methods of the production process from the users of products. Iconic forms also covered over the relations between experiential investigation, discovery and invention—this through what was objectified symbolically.

The projected truth claim of magic—what we would now classify as the first thinking in theoretical physics and the natural sciences of ancient civilizations, and the error of reasoning of the alchemist—were all, in fact, critical factors in filling the void between knowing nothing and knowing something. Ignorance was, and is, mediated by knowing in error. We err to know, frequently err when we do "know," and are often educated in error.

The whole metaphysical enterprise of Western thought can be seen to rest upon error, as it manifested blind faith in measurement and number. Such a reduction of truth to calculation hides the determinant power of the hermeneutics of error (the "learning" *that nevertheless occurs* by identification, interpretation and investigation from a mistaken proposition).

It is within this intertwining of truth and error that magic and alchemy need to be considered.

Alchemy is the meeting place of magic and science. It was not a consensual body of knowledge because it was not the product of one culture, one moment in time, a common objective or even of an orderly traffic in ideas. Alchemists claimed philosophers, physicians, astronomers and mystics among their number. Alchemy was an assemblage of secrets, the cryptic, the contradictory and the insightful. It was certainly not simply a flawed art aimed at turning base metals and other materials into gold. Rather it folds into grandiose and contradictory systems of philosophy that embody fields of human belief and ideas that extend across many cultures and over a vast expanse of time and space. Various histories can be traced.

We can identify at least four impure, partially inter-linked sources of what the West constituted as alchemy. There is an extremely old Indian tradition, as well as the Arabic (especially Babylonian, Persian, ancient Egyptian)—much of which was inscribed in a literature by the start of the –third century.[2] Then there is the Chinese formation of alchemy, with its strong connections to Taoism. The history of Chinese alchemic practice extends back 2,500 years and is linked to activities such as necromancy (divination via the spirits of the dead) and geomancy (divination via reading the "forces of nature," in particular feng [wind]-shui [water]). While working with metals, the Chinese tradition also focused upon finding an elixir of immortality. In contrast, Greek alchemy can be traced back to the Alexandrine era, seen in the age of the atomists and recognized in the thought of Plato and Aristotle.

Alchemic ideas arrived in Europe, most powerfully in the eleventh and twelfth centuries, via Latin translations of Arabic texts. The Greek concern was with the disclosure of the secrets of *physis* (which gets loosely translated as nature). While each of these bodies of knowledge had a corpus in its own right, they equally converged into new eclectic formations. So, for instance, we read of Jäbir ibn Hayyän, an Arab alchemist of great acclaim at the court of Harun ar Rashid, whose work was translated into Latin in the twelfth century. He employed a debased pre-Socratic theory of elements based on combining "the metals of the earth, under the influence of the planets, out of the union between sulfur and mercury."[3] Such traffic in ideas was at the core of the burgeoning of the Enlightenment, and the transformative character of its "ecology of mind." This is illustrated by the Western rediscovery of Greek thought which was translated (in the case of Aristotle's writings) from the Greek into Suriac, then Arabic, Hebrew, Latin and English. To a great degree, it was the translation work of Islamic scholars that delivered the knowledge of the ancients to the moderns and thereby created the possibility of the Western university.[4] In this "age of translation" peripatetic scholars moved from Moorish cultural centers in Spain to France, Germany and Italy, bringing knowledge of ancient Greeks with them. Until the twentieth century, Western knowledge largely concealed this Eastern tutelage—a concealment that was given great force by the mischaracterization of the Islamic world as barbaric and the attempt of the crusades to destroy it.

Alchemy was not separate from philosophy: this is evidenced in the significance of alchemy to the "schoolmen," that body of philosophers who led Europe out of the Dark Ages and were associated with the formation of the university—Albertus Magnus (1206–80), Roger Bacon (1214–92), St Thomas Aquinas (1227–74), Dun Scotus (1270–1308). At the same time, one of alchemy's key features across the traditions of China, Islamic science and the Greeks was that while practitioners had their own intellectual projects in the alchemic enterprise, they all employed practical arts that drew heavily on the kitchen and the laundry. Washing, drying, preserving, heating over a flame, baking, salting, boiling, etc. were all employed as tools of what was de facto, applied chemistry.

The practices of alchemy created a considerable empirical knowledge of metals and other materials, and of working with fire, which advanced specific knowledge of furnaces, fuels, the management and application of heat and thermochemical processes. Alchemy thus established one of the key foundations of modern chemistry and its methods. Yet it did so in its search for what we now know to be a scientific fool's errand. In this context, what has to be acknowledged is that a great deal was discovered in error.

To gain a true sense of the significance of alchemy, we need to look a little more closely at its history. While this will barely touch the vast literature of the topic, it will nevertheless give some weight to the argument of its importance.

Metallic life forces

Chinese alchemy was centered on a hylozoist proposition in which matter was viewed as being endowed with life forces—thus an elixir of life could be posited in the domain of metals. One of the ways in which this thinking expressed itself was to read color as a marker of relational connections. Thus, red was seen as a sign of life as it was the color of blood. In turn, this meant that hematite ore, a common and valued iron ore, and a much employed red pigment, was used to paint the body of the living and dead as a sign of its former life.[5] We also note that the very name of the ore was derived from the Greek word for blood.

Needham tells us that the first text of Chinese alchemy was most likely the *Shu Ching*, attributed to the –fifth century.[6] Its focus was on the creation of immortality by the development of practical techniques to transfer the "imperishable perfection" of gold to the human body: "… it was felt that somehow being could be transformed into a gold-like state." This state was to be attained via "some kind of potable gold."[7] Gold, as the material of perfection, was taken as a mark of "that without an end"—the medicine of immortality forming a direct connection to the Taoist cult of holy immortals.[8] Taoism, as the way to immortality, was profoundly influential as an ordering of knowledge. It assigned knowledge to "wizards and philosophers, the diviners and cosmological thinkers" but equally to the "practical domain gained by the hands and the direct observation of nature." Thus, the spiritual, the religious, magic, iconography, proto-science and the practical arts were all conflated. Alchemy was a framework which filtered the significance of "chemical observations."[9] What this meant was that a metallic elixir functioned in an entire regime of activities, which included dietary and gymnastic practices.

Eventually, the toxic dangers of the consumption of gold were recognized, although it is not exactly clear when.[10] However when it comes to iron it is a different story.

By the eleventh century, a whole book on the alchemical-pharmaceutical uses of iron was in circulation.[11] One of the "tonics" was to use ferric oxide by:

> … the rusting of plates of good steel in brine under controlled conditions, then administering it in a complex prescription with plant and other materials, so that small amounts of other more absorbable salts (such as citrate, malate or acetate) may be formed. The main product was called *thieh yin tan yin*, here meaning the "spontaneous successor" or "posterity" of iron.[12]

As a medical biologist, Needham comments that what resulted was "a tonic of the first order" that we now know would improve "haemoglobin levels and the erythrocyte count."[13]

Chinese alchemy contributed to proto-biological and chemical sciences with its development of sophisticated methods of classifying and describing plants, animals, minerals, chemical substances and their transformative states. Again, a ground of fact was established out of error. Effectively, an "affirmative empiricism of correction" (learning by doing and reflecting on it) travelled within practical arts, including the metallurgical.

Greek alchemic thought

Modes of thought do not simply come out of the blue; they have a historicity that may become a history. As far as Western thought is concerned, the history of how the Greeks constituted and transcended magic and myth is directly connected to how they became influential in the development of thinking towards reason and, within the orbit of our concerns, alchemy and metallurgy.

Speculation dominated early Greek thinking. As the history of all cultures evidences, without the invention and expansion of narratives that generate a rhetoric of ideas and categories, the power of observation remains restricted. Image and imaginaries, information and observation, constructions and concepts not only feed upon each other but require a considerable cultural sediment to function as knowledge. This is slowly built by an accumulation of naming, speculation, stories and myths on the forms, forces and appearances of the immediately encountered environment. Clearly, there was a vast expanse of time between the development of language, the creation of oral cultures, the creation of narratives that posited causality with "the gods" or other mythological forces and the arrival of those persons who became regarded as "the first thinkers." It is with such qualification that we approach the speculative thought of the earliest of the pre-Socratics. The questions these thinkers asked and the answers they postulated were of great tangibility and of immeasurable determinate consequences in setting the agenda out of which Western culture was formed. The questions were of the order of Thales asking "what held the earth up?" or of Anaxagoras calling for proof of the corporeality of air.[14] It was in such a context, between the −sixth and −fourth centuries, that a critique of the explanatory power of magic started to emerge. The focus of this critique was the Hippocratic writers' concern with the "sacred disease" (epilepsy), its cause and the claims of ritual cures (often based on the driving out of spirits).[15]

A key question posed in the historical literature on magic, called up by G. E. R. Lloyd, is—"can people with inefficient magical beliefs come to be critical of them, under what conditions and to what extent?"[16] One of the crucial factors in the discrediting of magic in pre-Socratic Greece was claimed to be simply a careful observation of circumstances. The claim was that Hippocratic writers gained authority by what they learnt from close description of epileptic fits. This view is contested by Lloyd.[17] He draws attention to the distinction between sustained observation and deliberate research (observations carried out to gain new data to test a hypothesis), and that both kinds of observation sit within a network of theoretical assumptions. He cites Frankfort, who said that as we might explain the breaking of a drought in terms of changes in atmospheric pressure bringing rain, the Babylonians observing the same downpour would attribute it to the intervention of a gigantic mythological bird. These comments make evident that a particular history of thinking, as expressed in language and manifested in an ecology of mind, is directive of the manner of seeing. As Plato made clear, we see with our minds—our eyes are mere instruments. Thus, we see what we know. Interpretation is therefore always predicated upon the historical experience (direct and mediated) and the pre-given values and meanings of the subject who seeks to know.

The passage from magic to science was not simply a move from fiction to fact or superstition to truth, but a shift in the object of belief. Both travel with faith. Alchemy thus can be viewed as a necessary transitional mechanism from faith in mythological forces which were unknown but believed in, to faith in the power of reasoning (with its concealment and negation of the unknown).

Alchemy was underscored by a theory of matter prefigured by pre-Socratic thinking about elemental forces. Most significant was a monist view of matter and world which postulated that all diversity is unified with the one—all difference is thus resolved in the singularity of this "one." Thales held that everything came from an originary element, Anaximander proclaimed the endless and the power of four qualities, and Anaximenes gave all causality to air, life and soul. Heraclites posited the agency of matter in fire, as the force of "natural" change. And the scientist/magician Empedocles was the first thinker to divide matter from force and to assert the permanency of process and the relation of elements. There was also the atomist tradition of Democritus and Epicurus, which viewed all change as the movement of particles of various sizes, shapes, positions, weight and dispositions.[18]

Against this backdrop emerged Aristotle, whose ideas held sway for centuries, in fact well into the Enlightenment. He held that change was driven hylomorphically, that is, by active matter with energy. His proposition was that all matter was reducible to the same matter, with variation coming from the differential presence of four elements: air, fire, earth and water, supplemented with four qualities: hot, dry, wet and cold. Alchemy joined this thinking to a metaphorics of the substances of sulfur and mercury, understood as "that fire that made solid" and "that solid that flowed." These materials were not viewed just as materiality: sulfur was the metaphor of combustibility, while mercury was a "liquid metal with the power to fuse." Matter was posited as an abstraction as well as an agent of transmutation. The ability to transmute was asserted to rest with "prima materia"—the matter from which all matter came and to which all matter could return.

While it is the best-known example of alchemy, the creation of gold by transmutation was not the limit case of change—the ultimate aim was in fact the realization of the "philosopher's stone," which was thought not only to enable base metals to change to gold but to transmute matter to the metaphysical (expressed as sophic sulfur, sophic mercury, spirit and soul). The philosopher's stone was characterized as the metaphysical materialized, with varied qualities—for instance, it was claimed to be red, heavy and sweet-smelling.[19]

Aristotle's theory of matter supported all those alchemic claims that base metal was fundamentally the same matter as gold. That this was incorrect, and the way it was incorrect, was one of the crucial factors that bound alchemy, metallurgy and chemistry together.[20] Metals in Aristotle's thought act as a relational link between his physics, astronomy and biology. Aitchison cites Aristotle thus: "… as one metal dies and another is born, the phenomenon is akin to the life cycle."[21] Hylozoism, as noted, was a significant alchemical idea that transposed the notion of matter being endowed with life to metals that lived, died, rotted, were reborn and multiplied. Moreover, this idea flowed into an alchemical monism that constituted life as the singularity of which all elements were elemental. The characterization of the elements of matter, life forms and heavenly bodies, and their structural relations became the basis for the construction of a symbolic language of alchemy that wove together number, music, image, color, form and mythology. This mystical symbology was appropriated, modified and transported into science and contemporary life in the psychology of Carl Jung by taking up Aristotle's exposition of *anima* and "the power of life forces."[22]

For all its error and mysticism, alchemy actually contributed a large slice of the qualitative prehistory of quantitative chemistry. George Starky is a figure who illustrates the ambiguous influence of alchemy in the rise of modern chemistry and scientific thought (be it with a good deal of concealment). The mid-seventeenth century marks the sunset of alchemy and sunrise of modern chemistry. We find Starky in this moment standing for a quasi-scientific community, for "patascience" and the influence of alchemic practice.

The story of an invisible man

To write about George Starky the alchemist requires we do so under his alchemical persona, Eirenaeous Philalethes—the elaborate alter ego he created and projected. Before we look at Philalethes, a biographic sketch of Starky will help us understand him in some kind of context—to do this we will draw on the account of his life by William Newman.[23]

Starky was born in Bermuda in 1628 of Scottish parents; his father was a Church of England minister. He was educated at Harvard (where he first became interested in alchemy), matriculating in 1643 and gaining a Master's degree in 1650. Prior to this, in 1648, he had started a medical practice, which had flourished. In 1650, Starky left New England for London. He did this because he wanted access to proper laboratory equipment, quality glass and an efficient furnace (a rare thing in his day); and he wished for the company of a scientific community of standing. A measure of his success on this count was that he became highly regarded by "the father of modern chemistry," Robert Boyle and that "giant" of scientific thought, Isaac Newton. Both of these men were influenced by his ideas and held him in high regard, not, however, as Starky, but as Philalethes.

Philalethes had a working relation with Boyle who financially supported him during 1651 and 1652 and viewed him as a skilled investigator of the "anatomy of metals." Some measure of the trust in which he was held is indicated by Boyle allowing Starky, the physician who quickly gained repute, to treat both himself and his family. The measure of the relationship is summarized by Newman: Boyle "… sought out the knowledge of Starky on matters chemical, philosophical, subsidized his experiments and submitted his own relations to Starky for Chemical cure."[24]

The influence upon Newton was less intimate and exclusively via the alter ego Philalethes.

Newton himself had been a practicing alchemist for many years. Clearly, this is at odds with his construction as an iconic figure in the history of science. In the 1680s, Newton compiled an index of his alchemical reading which ran to over 100 pages and listed 879 entries on alchemical topics. Philalethes made up 302 of these entries, over twice as many as any other author.[25] Newton's interest in him was specifically in relation to the "structure of matter" and chemical "affinity."[26] Also, Newton and Boyle corresponded on alchemic matters.

The influence of Philalethes extends beyond Boyle and Newton to a range of lesser figures, but also to others of eminence. These include Georg Ernst Stahl, a scientist of considerable importance and author of the *Philosophical Principles of Universal Chemistry* first published in 1723, and philosopher G. W. Leibniz, as evidenced in a letter to Adam Adamandus Kochanskia in Hannover in 1696.[27]

Alchemistry

These Enlightenment scientists did not see their concerns in terms of science versus non-science. It is only retrospectively that it became apparent that they were on the hinge between one mode of understanding matter and another. They viewed what they were doing as what has always been done—making new knowledge out of the insights and errors of the old. So understood, Boyle can be placed between an Aristotelian hylomorphism, "new science" and "modern mind" as represented by Galileo and Descartes, who viewed the world as a machine system of inert bodies moved by forces of physical necessity (this is the ecology of mind to which Newton contributed and from which many of his ideas stemmed—not least being gravity). Familiar knowledge for Boyle and his contemporaries was much nearer to alchemy than it was to the then still emergent science of chemistry.

As said, alchemy was not culturally uniform. In the moment of its transition to chemistry, its internal divisions become apparent. This was characterized as the activities of madcap "puffers" chasing the impossible goal of gold versus an emergent experimental science of matter that was starting to deliver useful results, especially in metallurgy and in medicine as "iatrochemistry" (alchemic chemistry in the service of medicine that was partly evident in the activities of Philalethes). Robert Boyle was a significant voice in the shift away from the quasi-science of error towards a new science on the nature of matter and practicality. His tract of 1661, *The Sceptical Chymist*, was a major intervention that turned alchemic culture towards contributing to what was to become modern chemistry (which arrived 100 years later).[28] Placing alchemy back in the account of the "ecology" of mind, in Chapter 1 we observe that Biringuccio devoted an entire chapter to alchemy in his *Pirotechnia*, presenting an insightful argument on alchemy and metallurgy that prefigured Boyle's shift from the mystical to the scientific by some 120 years. His views on alchemy were centuries ahead of their time and beg acknowledgement at some length. For example, Biringuccio writes:

> Besides the sweetness offered by the hope of one day possessing the rich goal that this art promises so liberally, it is surely a fine occupation, since in addition to being very useful to human need and convenience, it gives birth everyday to new and splendid effects such as the extraction of medicinal substances, colours and perfumes, and an infinite number of compositions of things. It is known that many arts have issued solely from it; indeed, without it or its means it would have been impossible for them ever to have been discovered by man except through divine revelation.
>
> Thus, in short, it can be said in conclusion that this art is the origin and foundation of many other arts, wherefore it should be held in reverence and practised. But he who practices must be ignorant neither of cause nor natural effect, and not too poor to support the expense. Neither should he do it from avarice, but only in order to enjoy the fine fruits of its effects and the knowledge of them, and that pleasing novelty which it shows to the experimenter in operation.
>
> The other path is very distant from this one, yet seems to have been born from it. Though it is sister or illegitimate daughter to it, it is called sophistic, violent, and unnatural. Usually only criminals and practitioners of fraud exercise it. It is an art founded only on appearance and show, one which corrupts the substances of metallic bodies with various poisonous mixtures and transforms them so greatly that it easily makes them appear at first sight to be what they are not. It often has the power of deceiving the judgement as well as the eye so that it appears beautiful to the one who has performed it, but later it is so much the more displeasing both to him and to all others when they see that its vestments fall and, when it stands revealed, they understand that it contains only vice, fraud, loss, fear, and shameful infamy.[29]

Alchemy: The contemporary account

Alchemy made a massive contribution to the experimental practices, documentation and progress of metallurgy. Irrespective of its primary goal, *en route* it increased knowledge of metals, minerals, ores, furnace technology and thermochemical processes. Its idealization of gold, the perfect material, and its totally flawed theory of materials, has obviously fallen by the wayside, yet the ambition of bringing into being materials with magical properties lives on. Today the ambition is more likely to be for materials that can endure in extreme environments of heat or cold; enable travel in hostile conditions on land, sea, air or space; or in a war setting, that can cause or withstand high impact. Metallic glass is one such contemporary material.

Metallic glass is an alloy created from many metals (iron, palladium, nickel, copper, titanium, aluminium and more). Unlike other metals it is amorphous, which means that as a result of ultra-rapid cooling at one mega kelvin per second it does not crystallize. It is this quality that makes it a glass (even though it is not transparent). The consequence of the rapid cooling is that the material's atoms, which exist in various sizes, are dispersed at random. It is this atomic condition of disorder that gives the material its character as the toughest material yet to be created. Needless to say it is extremely expensive.[30]

While being used for applications like golf club heads and wind turbine blades, unsurprisingly, this material is of special interest to the military (for armor-piercing shell heads, because rather than flattening out on impact like all other materials, it shears away and self-sharpens as it hits it target. Not surprisingly, its development has been partly funded by the US Army Research Office). Another area of super-advanced metals development, again driven by military applications, is that of carbon nanotube metal matrix composites. For example, the Japanese Type 10 MBT (Main Battle Tank) is built from a related material— triple hardened nano-crystal steel, combined with a modular ceramic composite armor.[31] Of course, the development of weapons has been indivisible from metallurgy from its inception.

The notion of a "wonder material" with magical properties has accompanied the introduction of many new materials in the modern age, most notably plastic.[32] Writing in 1957, at the height of the novelty of plastics, the celebrated French critic and cultural theorist Roland Barthes wrote that plastic is:

... in the essence the stuff of alchemy ... So more than a substance, plastic is the very idea of its infinite transformation ... And it is this fact, which makes it a miraculous substance: a miracle is always a sudden transformation of nature. Plastic remains impregnated throughout with this wonder: it is less a thing than a trace of movement ... The hierarchy of substance is abolished: a single one replaces them all: the whole world can be plasticized, and even life itself ...[33]

While there have been claims about various metals being "wonder materials"—most notably aluminium, to a lesser extent titanium, and now materials like magnesium and tantalum—these claims are not as highly charged as those made about plastic.[34] This is because plastic unleashed the notion of a synthetic universe that was not based on extracting materials from the earth and transforming their "nature," but rather, plastic held out the promise of the creation of a new nature by the harnessed power of chemicals; hence the euphoria and horror expressed by Barthes' essay.

Ecologies of the science of metals[35]

Metallurgy has played a role in both founding and advancing science. Retrospectively it can be claimed as a scientific practice prior to the creation of scientific knowledge. Cyril Stanley Smith points out in his introduction to Biringuccio's *Pirotechnia*:

> ... metallurgical methods that had been developed by trial and error prior to the seventeenth century were far ahead of chemical theory, and it was not until the eighteenth century that advances in the fundamental sciences had affected metallurgical practice to an extent that required the writing of entirely new books.[36]

As a domain of knowledge, metallurgy divides into two areas: process metallurgy, concerned with the extraction of metals from their ores, their reduction and refinement into workable materials; and physical metallurgy, concerned with the investigation and management of the properties and applications of metals in their production and use. Linked to physical metallurgy is metallography, which is concerned with the microscopic crystalline structures of metals.

One of the key Western figures in the shift of metallurgy from a practical art to a modern science was the eighteenth-century French aristocrat scientist Réné Antoine de Réaumur, whose seminal work of 1722, *L'Art de convertir le fer forgé en acier*, besides advancing process technology was the first study of the interior structure of the metals. From a contemporary perspective this central figure in the history of modern metallurgy was also a proto-ecologist, as evidenced by his six-volume entomological study, *L'Historie des insectes*, published between 1734 and 1742.

There are tens of thousands of metals and alloys. Furthermore, customized manufacture of alloyed metals means it is impossible to say just how many there are today. Notwithstanding this situation, and with particular reference to iron and steel, fundamental principles still apply. A brief review of these will contribute to the task at hand. While this will be only a very basic introduction to a technically complex subject, it does aim to make a number of connections that are generally absent in the scientific literature.

Process metallurgy: Ores, reduction and refining

The principal iron ores are oxides of iron occurring as the following minerals: magnetite (the richest ore with 65 percent of iron); hematite (a rich ore which ranges from 50 to 60 percent iron); goethite and limonite (hydrated forms of ferric oxide); taconite (a general name for iron-bearing rocks with 20 to 40 percent of iron); and jaspilite (a mixture of magnetite and hematite). There are also a number of sulfides of iron. Iron is also smelted from a carbonate ore—siderite (also called chalybeate). While the global distribution of ores is widespread, the geographic location of ore with high iron percentage is sparse. Those nations that made high-quality iron in the distant past, like Sweden, did so not because of superior knowledge and technology, but because of the availability of high-grade ore.

Because of the geographic variation in grades of iron ores, pre-treatment methods have been developed to assist smelting. The first method was the visual inspection and hand-picking of ore.[37] As the quantity of richest ores (which have ±60 percent of iron) diminish, it becomes necessary to increase the percentage of iron by a pre-treatment "beneficiation" processes. The most basic process is the washing of ores to remove lighter gangue (waste materials). Ores may also be crushed, with material separation occurring via gravity and flotation, or iron can be magnetically extracted from the gangue after crushing. Thereafter, the powdered iron has to be agglomerated by being sintered, a process invented almost a century ago, or turned into pelleted, nodualized or briquetted forms—a more expensive, complex process invented in the 1950s that has become increasingly important.[38] All of these processes are designed to increase the efficiency and capability of the chemical reactions of the blast furnace. Carbonate ores containing siderite are pre-treated by a calcining process that consists of heating the ore with hot air to drive off carbon dioxide and convert the carbonate element of the ore to oxide.

In the blast furnace method of making iron, once the ore has become part of the blast furnace burden (the sum of all material fed to the furnace), and the smelting process begun, it becomes transformed into metallic iron by a thermochemical process that changes as the material passes down through the temperature zones of the furnace stack. While the technology is basic, and fundamentally unchanged from its inception, the relation between materials and temperature, and the chemistry between materials, are extremely complex.[39] Depending on the make-up of the ore, initially small amounts of other materials get dissolved in the metal, most commonly silicon and manganese. While, in small percentages, these materials can be beneficial to the qualities of final product, conversely, phosphorus and sulfur are common contaminants. Both pig iron and blast furnace metal contain amounts of carbon partly dissolved in the metal as graphite. How carbon is held, and its percentage regulated, are key functions of the steel-making process. As well, the ceramic refractory alloying materials (silica, alumina, zirconia and magnesia) used to line furnaces are not chemically passive but interact with the acidic or basic properties of fluxes used in iron and steel-making, and have determinate consequences in the formation of the chemical characteristics of the slag as well as on the level of impurities in the metal.

In managing the thermochemistry of a conventional coke-fired blast furnace, the aim is to produce iron with as low a quantity of impurities as possible, and to minimize the expenditure of coke required to do this. Likewise, a major design factor in high capacity coke-fuelled and supplementary fuel-injected blast furnaces is dealing with the volume of the burden so as to create a large quantity of iron at a low cost. It is possible for charcoal to produce an iron low in silicon and carbon at much lower temperatures and with a cold blast, but as was pointed out in the previous chapter, only in much smaller quantities.

Iron can also be made by alternative means that do not require it to become molten for oxygen to be removed. The amount of iron made this way is becoming increasingly important, especially as an alternative feedstock to scrap for EAF production. Competing methods have been developed and while these have required large amounts of investment, there are significant operational cost advantages once initial development has taken place. These processes are having discernable consequence in the continued weakening of the structural position of blast furnace as the primary means of making.[40]

The physical metallurgy of steel: Crystalline structures

Primarily, the reasons for altering the physical qualities of steel have been to bring its performance into line with particular applications. Thus, if a material is required to be easily bent, the steel supplied needs to be ductile; when the steel has to withstand a pulling force, tensile strength is required; and when impact and compression strength are needed, a hard steel has to be employed. Needs can be more complex than just indicated; for instance, hardness may be required as the overall characteristic of the material or just its surface—this then becomes a matter of a secondary treatment. One old-established method of doing this was by case hardening (carburization) in which the metal's surface is brought into contact with carbon and a high temperature heat source (increasing the carbon content and thus the hardness of the surface). Sophisticated and specialist steels have been created in recent decades by more exotic alloy technology. Developments in armor plate for fighting vehicles (tanks, armored personnel carriers and armored cars) illustrate the point. Here the requirement for the plate is seemingly contradictory, yet ever open to being more adequately realized. The material needs to be as light as possible (so the vehicle can move quickly) while also being able to withstand the impact from high-speed projectiles, some of which have been specifically developed to pierce armor. Steel is being challenged in this area by new materials like carbon fiber, high-performance ceramic plate and composites that combine both of these materials—which are of the order of 20 percent of the weight of steel.[41]

Steel, as an alloy of carbon and iron, can be given an extremely wide band of characteristics just by varying the percentage of carbon, the temperature and method of cooling. Additionally, the qualities of steel can be significantly modified by alloying it with other metals. The crucial factor, as with carbon, is always the percentage used. The advancement of alloys obviously occupies a very significant position in the annals of metallurgy. Even elements like phosphorus, silicon and sulfur, which are usually regarded as contaminants, can bring positive attributes, such as machine-ability and tensile strength—but only if introduced in the correct amounts. Other metallic elements such as nickel, chromium, manganese, molybdenum and vanadium can be used as alloys to increase the hardness, toughness, tensile strength or corrosion resistance of steel.

Steel, in common with all metals, is made up of crystals that form in solidification. Once made, the chemical composition of the crystalline qualities remains constant. However, the particular physical qualities of the crystalline structure of steel—qualities that determine, for example, its strength, malleability, ductility, hardness, softness, toughness—can be changed by the management of heat or by working the steel, either hot or cold.

Although metallurgical issues will be looked at in later chapters, in the context of our comments here, it is worth looking at the composition of steel and its treatment in a little more detail.

In the crystalline structure of steel, carbon exists in two forms: in a solid solution in the iron (called *austinite*) and partly combined with the iron (called *cementite* or *iron carbide* that is hard and dark in color). Iron in which carbon has been dissolved is called *ferrite* (this is soft, ductile and light in color). The qualities of cementite and ferrite combine, in a laminar fashion, to form pearlite. So while steel is constituted as an alloy of iron and carbon, usually in a percentage range 0.20–1.0 percent carbon, the make-up of this percentage (and the relation between its solid and combined relations) determines much of the performative qualities of the material. Moreover, when steel is heated to temperatures over 900^{108}°C its atomic structure changes, dissolving all the carbon. How the metal is cooled and at what speed determines the structural reformations of carbon. If the steel is cooled "normally" in the open air there is no time for pearlite to form in anything but scattered forms (the results of this *normalizing* is a hard and tough steel). If the steel is cooled quickly by rapidly plunging it red-hot into a bath of oil or water (quenching) the formation of ferrite and pearlite is completely suppressed. The result is the retention of carbon in a state of imperfect solution (called *martensite*) and a hard brittle steel (that is not of any use in itself). However, once this steel is *tempered* by being reheated and then soaked at 500^{1080}°C, the carbon comes out of solution and reforms in a finely graded condition called *sorbite*—sorbite steel is a shock-resistant material. Another heat treatment that modifies the physical metallurgy of steel is annealing, of which there are various methods. The basic method takes steel up to a temperature above its critical range of carbon stability, holding it in this condition for several hours and then cooling it slowly. This results in a steel with a coarse pearlite structure that is highly machine-able.

The working of steel also alters its physical consequences: cold working/cold forming steel, after hot working and cleaning, increases its strength and hardness (but decreases its ductility) by distorting and lengthening the grain structure of the material. The cold worked steel remains fixed in its distorted condition, unable to recrystallize as hot worked steel does. The more hot steel is rolled the denser and tougher it becomes (a knowledge historically and empirically prefigured at the ancient forge, with the use of the forge hammer).

As we shall examine in more detail in a later chapter, historically, advances in metallurgy have been driven by the creation of the weapons of war. This holds true from the mythologies of a magical sword able to cut through any material. Mythical accounts recede back into the oral cultures of ancient peoples of East and West; they also extend out to the most advanced technologies of the modern war machine.

It was only in the nineteenth century that the impetus arose to standardize ways of making strong, tough and consistent quality steel. The drivers were the development of scientific metallurgy and the rapid expansion of the machine culture and economy of the Industrial Revolution. More specifically, the drive for standardized quality steel again came from the demand for modern armaments, be they light weapons like pistols, rifles and machine guns or heavy artillery and armor-plated war ships with long range naval guns. Progress was made by the adoption of more rigorous regimes of testing, this going hand in hand with the emergence of modern processes in engineering and industrial production. These developments link to the rise of America as the most technologically advanced nation and were manifested by, for example, the establishment of the ferrous metal testing program of the

Army Ordnance Department in 1841 and the establishment of the Navy Ordnance Bureau (along with the Federal Armories). These organizations formed key elements in the leadership of technical advances in the United States, and made major contributions to the creation and improvement of machine tools.

Science and modern metallurgy

As has already been touched on, the development of alloys to increase the performance of steel has been linked to metallurgical, military and industrial enterprise.

Clearly, the arrival of steel alloys vastly expanded the scope of the technical application of steel and the proliferation of products manufactured from it. This area of expertise gained a considerable dynamic from the mid-nineteenth century. It was during this period that manganese was added to create an improved tool steel; likewise, small amounts of nickel when added to medium carbon steel dramatically improved its toughness—molybdenum produced similar results. Another tool steel innovation was to add chromium to create a hard and long-life "file steel." The addition of tungsten was discovered to produce a very hard steel alloy ideal for bearings and cutting tools, then in 1900 a high percentage of tungsten (14–18 percent) was combined with a smaller amount of chromium (4 percent) to create "high-speed steel"—which gave birth to modern "self-hardening" machine tools. Another major innovation was, of course, the invention of the first stainless steel in 1914 by combining 14 percent chromium with medium carbon steel.[42] It is not our aim to review the now vast and complex world of alloys; there are, however, a few major points to be made in the context of the ecologies of steel.

The development of alloys and special materials has been driven by performative objectives. Such objectives are obviously technical and economic; however, they are now increasingly being viewed as environmental. Let's take two examples.

The first example is general and covers the implications of high-performance steel (HPS). These structural steels (starting with yield stress of typically 485Mpa) are placed in the range of standard structural grades (typically in the order of 400Mpa) through to advanced heat-treated alloy steels (typically in the order of 500–690Mpa). HPS is the name for a generic project that is being pursued and created by many steel corporations of many steel-making nations. The objective is to increase the strength of this structural material while reducing the volume of material used—by an order of 10 to 20 percent. Thus even if the cost of HPS is little higher, the overall cost of a structure would be cheaper. In bridge building, one of the key applications of HPS, besides reducing the amount of steel required, is the corresponding reduction in fabrication time, which is mainly by welding (weldability is a major factor in steel alloy metallurgy). Counter to the advantages of HPS, there are increased design demands, not least because it is harder to gain a rigid structure with less material. This requires the designer to turn away from familiar structural engineering practice and standards.

Notwithstanding design demands, HPS has the possibility of being a worthwhile enterprise not just because of economic advantages of high-performance reducing material volume but also because of the associated environmental gains. One of these is that using less steel in an all-steel structure will lower the embodied energy invested in it, which in turn means a lower level of attributed greenhouse gas emissions. Again, this presents design demands, for if a method of assembly is used that increases maintenance and disassembly time then the gains are dissipated.

Less material, and increased efficiency of production, does not of course automatically mean environmental gains.

For a material to do "more for less" and deliver environmental advantages, two prerequisites have to apply: first, the amount of energy used in the production of the material cannot be higher than the norm. Yet for HPS there is actually an increased energy to weight ratio—because the quenching and tempering to give the added performance requires extra energy (heating, cooling and handling are all energy-intensive).[43] The percentage of material reduction thus does not equal the percentage of environmental gain. Second, environmental gains only come from an overall reduction of the volume of material used—which cannot come only from intervening in material design and production, but also in extending the life of what is made (thus reducing the need for material replacement). Long life is clearly another way of thinking high performance; however, while this does in part depend upon the technically determined qualities of the material as well as how the components have been assembled and looked after, it also depends on the cultural value given to the object and its life. Currently, and obviously, two imperatives of the steel industry clash: its economics require an ever-increasing total volume of steel to be produced and sold; while global conditions of deepening unsustainability, in which the problems of global warming and climate change are implicated, create the counter-demand of reducing the environmental impact of steel production and application. In this context, if HPS just ends up adding one more product to a growing market for steel, then nothing is likely to be gained environmentally. On the other hand, if it does become a major substitute material, and make a contribution to an overall reduction in global steel production, then there will be real environmental gains.

The second example of a potential environmental benefit from a new material is a far more ambitious exercise—the creation of "ultra steels." This has been a multi-million dollar project of the Frontier Research Centre for Structural Materials of Japan's National Institute for Metals. About 100 research staff have been working to double the strength of 400MPa ferrite/pearlite steel by a process of micro-refining crystal grains to reduce their size by a factor of ten while also changing their geometry (specifically their grain-to-grain angle to an angle of at least $15^{108°}$).[44] This was the first step towards the ultimate aim which was to create a super-strong steel of 1,500MPa (carbon and manganese steels are typically 230–350Mpa, low alloys steels are typically 320–400Mpa and, as indicated above, advanced heat-treated alloy steels are typically in the order of 500–690Mpa). The same research center was also pursuing another "ultra steel" that will not corrode in sea water (which is well beyond the capability of stainless steel). For this to happen, a complex alloy experimentation program was designed and implemented.[45] As with alchemy, many such contemporary projects may be improbable quests, but as also with alchemy, what they yield in error may have profound consequences.

Notes

1   The exoteric indicates the symbolism of a closed body of doctrines of a "learned" body, while the esoteric means the secrets of magic, whim, mystical and arbitrary ideas. Alchemy created many books of secrets, like a series of books of 1532 called the *Kunstbüchlein* and the *Secreti* of Alessio Piemontese of 1555 (which was a compendium of information, recipes, and almost anything that the author deemed of interest).

2 The term alchemy has a clear association, its Arabic root—*al khem, alkimia*, meaning the art of the transmutation of metals and the knowledge thereby gained.

3 Leslie Aitchison *History of Metals* (Volume 1) London, Macdonald and Evans, 1960, p. 284, writes of Jäbir ibn Hayyän as a mysterious figure who created for himself, had created by others or who had a discipline with the Western name Geber (whose writings on alchemy became standard Western works).

4 The foundation of Western universities (e.g. Bologna 1088, Paris 1160, Oxford 1169) coincided with the period of the translation of a substantial body of texts from Arabic into Latin, including the first Western alchemical work in 1144. Ibid.

5 Joseph Needham *Science and Civilisation in China Volume 5, Chemistry and Chemical Technology*, Part 3 Cambridge: Cambridge University Press, 1976, pp. 3–6.

6 Many other texts were to appear in the period of the warring states (fourth to second century), see Joseph Needham op. cit., p. 4.

7 Ibid., p. 1.

8 Ibid., p. 6.

9 Ibid., p. 9.

10 Needham cites an eighth/ninth-century text, attributed to Chêng Yin, entitled *Chen Yuan Miao Tao Yao Lüeh* (translated as: Classified Essentials of the Mysterious Tao of the True Origins of Things), which mentions the dangers of 35 popular elixir formulae to life and health. Ibid., p. 20.

11 The *San Phin Shen Pao Ming Shen Tan Feng* (translated as: Efficacious Elixir Prescriptions of Three Grades Inducing the Appropriate Mentality for the Enterprise of Longevity), Joseph Needham *Science and Civilisation in China Volume 5, Chemistry and Chemical Technology*, Part 2 Cambridge: Cambridge University Press, 1976, p. 293.

12 Ibid.

13 Ibid.

14 G. E. R Lloyd writes with illumination and at length on this prehistory, see *Magic, Reason and Experience* London: Cambridge University Press, 1979.

15 For a detailed account of the significance of the "sacred disease," see ibid., pp. 15–58.

16 Ibid., p. 7.

17 Ibid., pp. 126–225.

18 John Dalton was to revisit and develop Greek atomic theory in the eighteenth century; one result of this was his theory of atomic weight.

19 John Read *The Alchemist in Life, Literature and Art* London: Thomas Nelson, 1947, p. 4.

20 Cutting across this seemingly rational progression was a trace of the magical, expressed as astrology. This asserted that the matter of the heavenly bodies was associated with particular metals, bodily organs and fates. An entire alchemical symbology was constructed around these relations. This symbology was to gain its most developed manifestation in European alchemy in the seventeenth century in a source of alchemical emblems (which carried the truths of the practice)—*Michael Maier's Alalanta Fugiens* (reproduction, edited by H. M. E De Jong, Leiden: E. J. Brill, 1969). Michael Maier wrote a number of books that reviewed alchemy, with the aim of elevating the few and damning the mass of cheats and impostors. He concluded that alchemy could be a serious study of matter, creation, God and his creatures, viewing it as a sacred science in which the liberal arts, medicine, mineralogy, philosophy and theology could all meet.

21 Aitchison *History of Metals*, pp. 226–7.

22 See Read *The Alchemist in Life, Literature and Art*.

23 William R. Newman *Gehennical Fire: The Lives of George Starky, an American Alchemist in the Scientific Revolution* Cambridge, MA: Harvard University Press, 1994.

24 Ibid., p. 52.

25 Ibid., pp. 228–9.

26 Ibid., p. 228.

27 Ibid., p. 209 and p. 279 n. 10.

28 Read *The Alchemist in Life, Literature and Art*, pp. 7–8.

29 Vannoccio Biringuccio *Pirotechnia* Cambridge, MA: The MIT Press, 1966, p. 337.

30 Dave Thier "What the Heck is Metallic Glass," *Forbes*, April 20, 2012, p. 3.

31 S. R. Bakshi, D. Lahiri and A. Argawal. "Carbon nanotube reinforced metal matric composites: a review" in *International Materials Review*, Volume 55, 2010, p. 41.

32 On plastic, see especially Jeffrey Meikle *American Plastic: A Cultural History* New Brunswick, NJ: Rutgers University Press, 1995.

33 Roland Barthes "Plastic" *Mythologies* (trans. Annette Lavers of the 1957 French edition) London: Paladin, 1973, pp. 97–9.

34 The claim for magnesium being a "wonder material of the new millennium" is that it is very lightweight, extremely strong, abundant and easily mined; but has a major downside—it takes an enormous amount of energy to make, far more than aluminium, which itself is very energy-intensive. See T. M. Pollock, "Weight Loss with Magnesium Alloys" *Science* Volume 328 No. 5981, May 21, 2010, p. 986. Tantalum is a rare metal with excellent conductive properties used extensively in the electronics industry, with almost two-thirds of the world's tantalum going to produce high-quality capacitors for mobile phones, laptop computers and other electronic devices. Its sourcing has been controversial—with large amounts of cheap supplies coming from illegal mines controlled by rebel groups in the Democratic Republic of Congo, with access to the ore ("coltan") and its trade becoming a stake in a 10-year civil war that resulted in the death of over 5.4 million Congolese. See Stephen Hutcheon, "Out of Africa: The blood Tantalum in your mobile phone" http://www.smh.com.au/articles/2009/05/08/1241289162634.html. Also see "Chemistry in Its Element—Tantalum" *Chemistry World*, Royal Society of Chemistry http://www.rsc.org/chemistryworld/podcast/interactive_periodic_table_transcripts/tantalum.asp

35   The following has been drawn from D. J. O. Brandt *The Manufacture of Iron and Steel* London: The English Universities Press, 1960, Norman J. G. Pounds *The Geography of Iron and Steel* London: Hutchinson, 1959, and BHP *The Making of Iron and Steel* Melbourne: BHP Steel, 1998.

36   Biringuccio *Pirotechnia*, p. xix.

37   Andrew Ure's *Dictionary of Arts, Manufacture and Mines* Volume 2, London: Longman, Brown, Green and Longmans, 1853—one of the most celebrated reference works of its age—has a 16-page entry on metallurgy; in describing "preparation of ores for the smelting house" Ure writes: "There is for the most part a building erected near the output of the mine, in which the breaking and picking of ores are performed. In a covered gallery, or under a shed, banks of earth are thrown up, and divided into separate beds, on which a thick plate of cast iron is laid. On this plate, elderly workmen, women and children, break the ore with hand hammers, then pick and sort them piece by piece" (p. 146).

38   In sintering, powdered iron and flakes of iron oxide (mill scale) are mixed with a flux binder like limestone, dolomite, quartzite and serpentine, coke breeze (small particles of coke retained in a screening plant) and baked. This is so it may be introduced into the blast furnace as a load-bearing material able to facilitate the passage of air.

39   First of all iron ore is reduced with carbon (coke)—the carbon being oxided to form the reducing agent carbon monoxide (which is extracted). Direct reduction consists of four *exothermic* reactions: first ferrosoferric oxide and carbon dioxide are produced, this then is reduced to ferrous oxide and carbon dioxide; then ferrous oxide and carbon monoxide are created, and finally liquid iron arrives as it dissolves carbon. This total reaction absorbs heat and is complete at temperatures over 800°C. To create this heat, two-thirds of the coke (carbon) burden of the furnace is combusted with the assistance of the air blast. The remaining one-third is used in the reduction process itself as the carbon extracts oxygen from the ore. As indicated, the ability of iron to melt is dependent upon the absorption of carbon. During this reduction process, the gangue in the ore joins with the flux material to form a fluid slag which drains from the furnace. The reduction process combines with an oxidization process as materials pass down through the furnace. By the time they reach the bosh (the lower part of the furnace opposite the tuyères), where the blast is the hottest, the temperature will have reached 1,800°C and the coke will have been fully combusted.

40 One of the dominant alternative iron-making methods is *direct reduction* using either coal or natural gas. This can produce iron in a variety of forms, for example: sponge iron, commercially marketed as "direct reduced iron" (DRI); hot briquetted iron (HBI—which is in fact DRI turned into a more marketable form); and iron carbide. Direct reduction delivers iron in a solid form able to be introduced into electric arc furnaces that are only able to melt, rather than smelt, metals. Direct reduction iron utilizes either natural gas (with reactors to manage the chemistry as in the FINMET and MIDREX processes) or coal (mostly employing a rotary kiln furnace as in the FASTMET process or the HIsmelt rapid process). Another alternative to the blast furnace is the use of a high temperature smelting reactor, as with the COREX process in which coal and iron ore are directly introduced into a bath of molten metal and slag. Again, the chemical process is based on carbon monoxide as a reducing agent. Direct reduction processes are becoming faster and delivering significant output.

41 This type of material has been used for the US Marine-commissioned fast and lightweight Reconnaissance, Surveillance and Targeting Vehicle (referred to as the RST-V)—see Adam Marcus "Robojeep" *New Scientist* June 13, 1998, pp. 36–7.

42 See Aitchison *History of Metals* on this topic.

43 Brian Fortner *Civil Engineering* April 1999, p. 61.

44 Furukawa Tsukasa "Japan's Search for Ultra Steel" *Iron-Age New-Steel* Volume 4, No. 3, 1998, pp. 76–80.

45 Ibid.

PART TWO

THE AGES OF INDUSTRY

FIGURE 4.1 Molten steel pour at Jones and Laughlin Steel Company, Pittsburgh, May 1942. Courtesy of US National Archives and Records Administration

4 THE PROTO-MODERN

This part of the book deals with the relation between steel and modernity. There are three large problems in dealing with such a topic.

First, while we can construct a complex narrative of steel-making, its use and world-transformative consequences, its actual impacts will always be greater than any account can convey. The ubiquity of steel makes it difficult for us to perceive just how deeply embedded it is in the infrastructure of the modern world.

Second, modernity itself is complex and not able to be reduced to a single moment or object of consideration. Moreover, modernity's "darker side" is increasingly being acknowledged and studied within the frame of post-colonialism and decolonialization.[1] Thus it is quite inappropriate to backload a unified historical picture onto the differences of those projects that set out to make: the modern global economy; modern science, technology and industries; the modern built and organizational environments; the modern mind and subject; modern political philosophy, institutions and nation states; and of course modern everyday life.

Third, the arrival of the modern has not been chronologically uniform. There has never been a single moment of modern time—the reverse is the case. The more time that has passed since the initial efforts to create the modern, the more that modern reality has fragmented. Thus, in the contemporary world we see nations where "the modern" is deemed as the past (the claim of "post-industrial" nations with a "postmodern" culture). In contrast, there are nations in which forms of modernity arrived only partially, were never fully secured and have now reverted to a dysfunctional version of their past (for example, those neo-tribal nations of Africa without a functional state or economy). And then there are those nations with rapidly modernizing industrial infrastructure, functioning in a global economy, but whose culture is of another, pre-modern lifeworld, or a hybrid of modern and pre-modern. Finally, there are those nations that strive to operate in the modern world economically and thereby gain material benefits, while being committed to anti-modern political ideologies or theologies.

Our approach cannot surmount this complexity, but it can allow the complexity to show itself. What all of the chapters of this part of the book share is the proposition that materials like steel, its making and uses have played a significant part in the creation of a technically infused ecology, which was first seen with the proto-modern, grew to a fully mechanized modernity (led by the USA and the advanced economies of Europe), and then finally flowed into projects and economies that assert themselves as "after the modern."

The history of steel-making is one of constant reinvention and innovation. Change is sometimes fast, but very often slow, and frequently does not follow a linear evolutionary path. The drivers of change are many and varied—downward market price, pressures on materials and labor, demands for greater volume and faster supply of material, technically generated needs for higher performance and socio-political pressure on corporations to secure less environmentally damaging processes. There are, however, other less pragmatic and more political drivers, like those framed by the agenda of economic and cultural modernity. In this context, steel has been deployed as a sign of modern nationhood for several centuries.

The ascendancy and deployment of EAF steel-making linked to the mini-mill is an example of how pragmatic and politico-ideological drivers are conjoined. Steel industry innovation and expansionism created this technology. But what it also allowed was for nations to buy into, and symbolically mobilize, steel-making at a much lower cost, because (as discussed in Chapter 3) the EAF/mini-mill configuration is a fraction of the vast investment required to build an integrated steel works. While EAF technology is not without problems, it continues to have significant consequences in transforming the steel industry.[2] Conceptually, EAF technology reinforces the message that steel is a material able to be conserved in use for reuse.

Once steel and modernity are viewed in relation to each other, a number of thematics become possible to explore. Of particular interest to us will be those material technologies and forces of labor that extended the use of steel in modern manufacturing. Before dealing with this, a historical sketch, some reiteration and commentary may prove helpful, especially to those readers who are less familiar with the history of the industry.

Iron: Historical snapshot

In many areas of human endeavor in Europe between the fall of Rome and the tenth century (the Dark Ages), as much knowledge was lost as gained. Iron-making was an example of this. It remained almost unchanged for many centuries and was carried out by smelting ore in inefficient charcoal-burning hearth-type bloomery furnaces that varied little from those used by the Romans. However, the number of furnaces started to significantly grow in the twelfth and thirteenth centuries—with, as noted in an earlier chapter, deforestation consequences. These furnaces could not create liquid iron, rather they produced blooms—pasty lumps of iron that had to be worked at the forge, or by a forge hammer, to render the iron into usability (working iron in this way improved the structural quality of the material and partly de-slagged it).

Small bloomery furnaces were increasingly displaced in the more advanced iron-making nations of Europe during the thirteenth and fourteenth centuries. In France and Spain, this was by the larger "Catalan forge," which by the sixteenth century could introduce a blast via a large-scale air aspirator (called a "trompe"). During the same period in Northern Europe, and especially in Sweden, the Osmund iron-smelting furnace was established. This shaft furnace was a significant factor in Sweden gaining its long-time reputation as a quality iron producer. From the Dark Ages and during the course of the Middle Ages, England continued to make iron by outdated and backward means.

By the fifteenth century, more efficient shaft furnaces were starting to be used more generally in Europe—it was starting to catch up with technological developments that often had been long established elsewhere. The shaft furnace was part of the history of ancient iron-making in the Middle East, Asia and Africa.

The European literature on the history of iron-making problematically elevated the German Stückofen furnace as the paradigmatic example of this technology. However, as acknowledged, other blast furnace technology pre-dated it. What is uncontested though is that this type of furnace marked a turning point in steel-making. It enabled a shift from the limits of pasty bloomery iron to liquid iron cast in "pigs," which became the industry base material. This became possible once the height of the furnace was increased and a constant cold air blast was introduced into the tuyères from water-driven bellows.[3] Modern furnace design, construction improvements, performance, size, fuel use, blast and so on were all built on the technological foundation of this earlier technology. The shaft furnace, in its developed form, enabled liquid reduced iron to drop down through a zone of charcoal, absorbing carbon en route. Carburized iron, with 3 to 4 percent carbon, then flowed to the base of the furnace, with the non-metallic materials now mostly left in the slag.

Steel and the first machine age

The shortage of charcoal (resulting from decimation of forests) gave a great impetus to finding a way to use coal to make iron. The European solution was to produce an almost pure form of carbon—coke. This development was generative of technical and industrial changes that had profound consequences. These events centered on Abraham Darby's iron works at Coalbrookdale, in Shropshire, England, where iron was first smelted with coke in 1709.

Coalbrookdale is often cited as one of the birthplaces of the industrial age, as it burgeoned into the first machine age by the mid-eighteenth century. Such a characterization undercuts a wider and more complex frame of reference.

Modern industrial tools, machines and production were all made possible by advances in metallurgy. More specifically, every advance in metallurgy created a corresponding advance in engineering. However, this did not happen as a result of a break with pre-modern technologies, but the reverse—these were frequently appropriated. The waterwheel is a good example.

Staying with the contribution of Abraham Darby, Howard Roepke writes:

Darby continued for many years to be preoccupied with the problem of increasing the power of the blast. A couple decades after the original success, the younger Darby was one of the early users of atmospheric steam engines to pump water into higher ponds. This enabled him to use 24-foot water wheels to operate the largest pair of bellows that had ever been made.[4]

Here not only is a clear example of the appropriation of a pre-modern technology, but one developed for the same ends over 1,500 years prior in China.

The waterwheel-powered bellows was superseded by a blowing machine, albeit in the modern age, driven by steam. The idea of powering bellows by a waterwheel arose in different circumstances and at very different historical moments in the technological advance of many nations—it is a good example of a situated, rather than evolutionary, pattern of development. Certainly the claim that Derby had built the largest bellows ever is contestable. The notion of a singular "progress over time" is also confounded on the small as well as the large stage of history. A report submitted by a M. Auguste Perdonnet to a French society of science and industry in 1831, cited by Harry Scrivenor, observes the poor state of French iron works and their failure to follow the logic of "the progress of improvements." Perdonnet's remarks illustrate the narrative inscription of technological evolution, as he notes: "we yet find at a great number of the water wheels, wretchedly constructed, and blowing machines more miserable."[5]

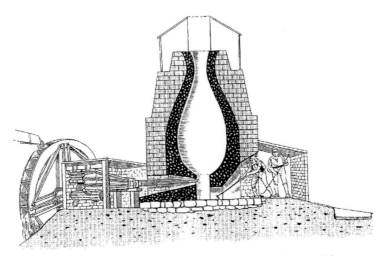

FIGURE 4.2 A European blast furnace as described in Biringuccio's *Pirotechnia* of 1539. Source: Vannoccio Biringuccio *Pirotechnia* Cambridge, MA: The MIT Press, 1966 (reissue of the 1942 edition)

A change of scale

The conversion of iron to steel in volume had to wait until the arrival of industrial methods that united modern metallurgical knowledge with large-scale systematized technological processes. This happened in the mid-nineteenth century—and two methods dominated: the Bessemer converter and the Siemens-Martin open-hearth furnace process. The latter was expensive to operate, rather slow, able to run on charges of pig iron and scrap, and produced steel in large volumes. In contrast, the Bessemer steel-making furnace was very fast, could only use pig iron, produced steel in small volumes and was harder to quality control.

The initial converter developed by Henry Bessemer in 1856 could blast atmospheric air to decarburize grey pig iron and turn it into malleable iron. It did this by a blast from the base of the egg-shaped converter (lined with a refractory material) passing through the molten iron which had been poured into it. Carbon and other impurities were then oxidized and expelled as gas and slag. The heat generated by chemical changes kept the metal molten. What was accidentally overlooked by Bessemer, as a result of unknowingly using low phosphorus ore, was that the process was unable to remove this impurity. This failing damaged Bessemer's reputation and set back the introduction of the technology by over 20 years. The problem was solved by the introduction of lime to create an alkaline slag that absorbs the phosphorus. However, there was a problem with the alkali reacting with the silica refractory material at high temperatures and destroying it. This was solved by Sidney and Percy Gilchrist, who added crushed dolomite to the refractory brick, giving it an alkaline reaction.[6] Effectively, the Bessemer converter could be used to regulate the iron/carbon ratio—hence its success as a steel-making technology.

At almost the same time as the Bessemer converter was introduced, the open-hearth furnace was being (re)invented by Siemens-Martin. In contrast to the 12-minute Bessemer steel-making process, which was so fast that it was extremely hard to control, the open-hearth furnace, while more manageable, took almost a day to produce steel.[7]

As seen, Darby's ability to make pig iron with coke in the first decade of the eighteenth century was a crucial development which went well beyond just solving the problem of replacing charcoal. Coke's load-bearing and material qualities (especially its low sulfur content and high carbon purity) enabled the construction of a larger furnace able

FIGURE 4.3 Bessemer Converter. Nineteenth-century engraving. Source: Robert Cochrane *The Romance of Industry and Invention* Philadelphia, PA: J. B. Lippincott Company, 1897

to operate with a more powerful blast and at higher temperatures, which in turn combusted contaminated material. Coke was a key factor in progressing new furnace technology. Its introduction led to a growing understanding and ability to manage its multi-function as a fuel, carbon source and reduction element in blast furnace chemistry. Specific innovations in furnace technology were also significant, and continued throughout the nineteenth century. Glasgow's gasworks engineer James Neilson, for instance, devised a way to introduce pre-heated air into a blast furnace (via the use of a coal-fired stove) in 1828. This meant that actual furnace energy was no longer needed to heat the inducted cold air of the blast, resulting in greater furnace efficiency, a more stable furnace temperature and the ability to use lower grade fuel. This innovation led to many more technological developments resulting in the pre-heating of furnaces using waste gases (already by 1854 Andrew Ure was reviewing many methods of doing this in the fourth edition of his *Dictionary of Arts, Manufacturing and Mines*).[8] As these advances were refined, the regenerative principle was established. This meant that hot furnace gases could be drawn back through flues to heat them; this prior to them in turn heating the air blast—once the flues had cooled, the process was repeated. This principle was a key feature of the Siemens-Martin open-hearth furnace.

An expanding iron and steel industry was powering metallurgical research from the eighteenth century onward. Here is the context in which we can view the work of French metallurgist René-Antoine Ferchault de Réaumur whose research (1722) led to a more malleable and tougher cast iron. Likewise Henry Cort's puddling process (patented in 1784) of making steel in a reverberatory furnace—by stirring the iron to constantly expose it to direct flame and thus subject it to an oxidization process (with the addition of iron oxides) which uniformly decarburized the iron—was another major advance.[9] Earlier, Benjamin Huntsman's ability to make good "cast engineering tool steel" in the 1730s contributed to the productivity and quality of light engineering (by providing improved steel for cutting tools).

Conversely, demands from new industries, manufacturing processes and technologies led to improvements in iron and steel production and products.

At this juncture, an increasingly complex inter-relation of science, technology, knowledge, materials, production processes and products become clearly visible. It is in this setting that we should view, for example, the coming of James Watt's, and then Boulton and Watt's steam engines, or more specific developments like the Nasmyth steam forge and Smeaton's casting of large cylinders for very large blowing machines required by "modern" blast furnaces. This history of invention and innovation was linked to the rise of what was to become the most powerful empire (Britain) and then the most powerful economy (United States) the world had ever seen.

Technological advances dramatically increased the mechanization of "labor power," which not only underpinned the expansionary force of the material "inputs and output" that underscored the second industrial age—the machine age—but also generated environmental and social impacts. Moreover, these developments also marked a change in the nature of time. This occurred through the circular dynamic of an ever-faster speed of industrial output that led to a speed-up of much else: life as lived; raw material utilization; increasing volumes of production; the movement and global distribution of goods; the exchange of money; the traffic of information; and the division between the colonizers and the colonized. Iron and steel were part of the very fabric of these changes, and to simply view them as materials is to abstract them from the complex environment from whence they came and from the ecologies of which they were elementally formative.[10]

China update: Other developments and the modern

Before continuing with this narrative gloss on the modem, it is worth again briefly looking elsewhere to provide a perspective that allows us to view this dominantly European history somewhat differently.

Our earlier account of iron-making in early China sought to show that there is no unified narrative of the history of iron and steel or, even more problematically, of their "development." Taking this further, it can be observed that the very notion of what is taken to be "the developed" has been a construct of modernity that was mobilized against other pathways to futures. "Development" became inscribed in economic and political theory on the presumption that a modern capitalist economy and culture was the future that all peoples in all places should have. Modernization, and latterly globalization, has been advanced by specific agencies that took such an understanding of "development" as logic to legitimize their actions. The endeavor to establish the hegemony of modernity was based not so much on an overt project or a clearly expressed political ideology but rather, and more fundamentally, on thinking that became generalized and embedded in the "ecology of mind" of the West. Over an extensive period this constructed way of thinking became established and enacted as "common sense"—infused in everything from high culture and national policy to the attitudes expressed in everyday conversation at all levels of Western society. As Antonio Gramsci argued, what we are here calling "common sense" is the foundational nature of ideology (which is always present, active and unquestioned rather than being something doctrinaire or concealed).[11] From this frame of reference we can see an unbroken sequence which links, for instance, mercantile capitalism, the ethnocidal violence of colonialism, the politically administered regimes of colonial domination, the neo-economic and cultural colonialism of the 1950s and 60s, the linking of aid to "development" in the "soft" coercion of world trade agreements and World Bank loans.

Momentarily resisting our own induction into the "logic of development," we need to realize that other forms of development were able to be imagined; however, no other path, no other notion of "progress" has been tolerated by modernity's dominant political and economic order.

Putting China into this picture, we can note that the advanced organizational structures of its civil society were seriously damaged during the Enlightenment by a colonial model of development that designated the Chinese as undeveloped and heathen. Lack of technological advancement was regarded as a developmental failure. Notwithstanding those debates on the nature of Chinese science and associated advancements in technology, or the lack of a shift from applied to experimental science, European observers mostly failed to see the enormity of what had been developed in China. The Eurocentric colonizing sensibility deemed the difference of others as evidence of deficiency. Nothing of the other was valued, or in many cases even acknowledged—as so many histories of colonialism now show, development demanded *the creation* of a condition of *under*development, rendering the cultural and economic fabric of others either irrelevant or dysfunctional.

Notwithstanding a long history of iron and steel-making in China, European colonizers gave almost no recognition or value to local knowledge, skills or resources. By the last two decades of the eighteenth century, iron and steel were being exported to China by the French, Dutch, English and Swedes. Immediately prior to this, China had been used to dump the material when demand was low. The volume of material traded steadily increased over the nineteenth century, and this acted as one of the key triggers in China's industrialization. Interestingly, by the end of the century about half of the 112,000 tons of imported metal was scrap.[12] Trade with China was very much on Europe's terms, especially after its defeat in the first Opium War. The 1842 Treaty of Nanjing formally inscribed the pattern of unequal relations. The local iron industry—already in decline—was seriously damaged by these events. In a country with rich ore deposits, as Warner argues, the decline was due to outdated technology, cheap imports and lack of investment capital.[13] The first sign of a reversal in the situation came in 1891 with the construction of a modern iron works at Hanyang and Hubie; however, it was the First World War that really changed circumstances.

The demand for iron generated by the war increased its price on the world market. This returned China to the fold of iron producer, even though to produce the tonnage required meant bringing back into use large numbers of what had been thought to be redundant traditional small-scale blast furnaces.

While there was a continual process of industrialization from the end of the nineteenth century, and while modern China was producing iron throughout the twentieth century, with some significant scientific advancements made in the 1930s, it was not until the revolution of 1949 that the demand for iron and steel-making dramatically increased. From this moment onward, the industry was projected as a sign of modern industrial nationhood. To meet these material and symbolic demands many small traditional steel works were created. The scale of this activity significantly changed with the "Great Leap Forward" program of communist China during 1958–9. This campaign elevated iron and steel as the iconic materials marking modern China, and they have remained in this position ever since. The initial national objective was to dramatically increase output. Thousands of backyard furnaces were built. Fifty million Chinese peasants were drafted into the iron and steel industry almost overnight. What they produced was effectively ideological iron—progress was measured in tonnage. Optimistically, around 4 to 5 million tons of pig iron was problematically, but officially, claimed as usable. However, as Warner says: "that is, 30 percent of the year's total pig iron production (13.69 million tons of usable pig iron was produced in these primitive blast furnaces) which, *in the opinion of most observers* was totally worthless" (our emphasis). The whole project was a "fiasco." The reaction to this experience played a major part in changing the mood, policy, action and expenditure of the national iron and steel industry during the 1960s. It was from this moment that China set out on the path to become the world's largest maker and supplier of the material: a goal that it achieved by the 1990s.

The modernization of labor and the machine

We are now going to look at a particular area of technology more specifically, and with a sharper focus. This will involve working through another kind of history—one that links material technologies, labor power, steel and modern manufacturing. All of these areas mark a very strong link between design and innovation. But in both cases these actions were intrinsic to applied practices.

As argued, the growth of the steel industry and its economic, socio-cultural and environmental impact on the making of the modern world cannot be understood by just focusing on the material itself. Between the material and the mass of products made from it, lie the force of its transformation as labor power, means of production and mental labor that became embodied in machine tools and organized labor processes, as well as becoming inscribed into the very bodies of a labor force. We are going to look at this at some length via three perspectives: the human body as site of impacts; the mechanization of labor; and the rise of machine tools. However, this history of the transformation of labor power and its relation to the human body started long before the modern machine age.

The human body as site of impacts

Biringuccio's observations in his *Pirotechnia* can be taken to stand for the skills and judgment of much earlier times. On writing on the art of the "smith who works in iron," he says:

> … even limiting the use of hammers [i.e., smithy work] to working of iron alone, it seems to me that beyond comparison it has more uses and that it has more secrets and perhaps more ingenious secrets than the art of any other metal. Thus, if it were not an activity so laborious and without delicacy, I would say that it is one greatly to be lauded, for when I consider that the master works without moulds or pattern, letting only the eye and good judgement suffice for it, and that they make exact and good shape by hammering alone, it seems to me a great thing.[14]

Equally, we can take Biringuccio's observation forward into the early machine age to the practices of the metal trades. Here, for instance, we find steelworkers bringing a disposition that connected back to the knowledge vested in the very earliest moments of iron-making to the development of an industry and its technologies that was bent on the erasure of dependence on skills of craft workers. As Robert Gordon comments in a caption illustrating an early Bessemer converter:

"A nineteenth-century blower had no instruments to indicate the progress of the conversion reaction; instead, he controlled the process by his interpretation of the appearance of the flame."[15] The propensity of industry to establish processes to disembody skill, and to inscribe these in technology, was of course a general trend in managerial and economic control of labor power. And it certainly was not just a feature of the operation of the steel industry. Again, we are in the same realm of "common sense" as we were with the ideological functionality of modernity's logic of "development."

The appropriation of skill to the machine which occurred in the second industrial age did not instantly mean that the everyday physical pain of working in the iron and steel industry simply vanished for large numbers of workers. The hardship and physical suffering of iron and steel workers was protracted. This common suffering was one of the reasons why the culture, and the associated social ecology of the industry, was so strong, and why its labor organizations became so important in changing workers' conditions. This history is a corrective to those seemingly concrete histories of the iron and steel industry that are actually depopulated accounts of technical change. To present a narrative without acknowledging the actions performed by, and the suffering of, live labor is in some ways an extension of the injustice of the times that turned a "blind eye" to such suffering.

To illustrate, let's take the case of Cort's puddling process, which often gets cited as a significant, but fuel-extravagant method of steel-making established in the late eighteenth century. Pig iron was placed in a furnace, which the puddler accessed through a hole in the side, using an 8-foot long iron bar to work the increasingly viscous iron into several 200-lb balls (440kg). The puddled balls of iron were then lifted with tongs suspended from an overhead track into a buggy to take then to the steam hammer or "squeezer" where more of the slag was worked out. Puddling required judgment and skill as well as strength. The puddler was constantly reading the changing condition of the iron and responding by adjusting heat flow and stirring action; timing was crucial, especially knowing when the iron was ready to be balled, which happened when boiling had subsided, the volume of slag diminished and the iron began to form clusters—a condition the puddler referred to as "the iron having come to nature."[16]

Of the process, Norman Pound writes: "Puddling has been described as the heaviest form of labour ever regularly undertaken by man. The

FIGURE 4.4 Leather aprons and gloves, steel boots—the protective clothing of early twentieth-century steel-men. Photographer unknown. Source: Norman J. G. Pounds *The Geography of Iron and Steel* London: Hutchinson and Co., 1959

physical effort required to manipulate the ball of heavy metal, accompanied by the heat, smoke and glare of the furnace, made puddling indeed a feat of endurance."[17] Commenting on the process in 1918, which was some 130 years after its introduction, F. W. Harbord and J. W. Hall write that it required: "… severe physical exertion on the part of the workman, the success of the process depending largely on his physical powers, which experience proves are not sufficient to enable him to deal efficiently with more than about five cwt of metal at a time."[18] These two observations, both in moderate language, create an image of extraordinarily hard labor that is beyond the realm of experience and imagination of post-industrial classes. Other examples are easy to find, like the work of the nineteenth-century "melter"—whose job was to tend the 60-pound (132kg) crucible and to decide the precise moment to extract it from the furnace with tongs and in one rapid movement pour it. This man worked the entire shift with his arms and legs wrapped in wet rags for protection from the heat. The work of the shingler was just as hard as that of the puddler and the melter. Again with tongs, held at arms' length, this man brought a lump of white-hot metal out of the forge to the violent action and deafening noise of a massive steam forge hammer, which worked the metal and turned it into bar. In order to do this his feet and legs were encased in armor, his body wrapped in leather and his head covered with a gauze mask to protect him from flying metal shards of scale and sparks.

The making of crucible steel was similarly energy- and skilled labor-intensive. The men who pulled the pots of molten steel from the furnace and poured them into molds (using long tongs like a puddler) had to wrap their legs in multiple layers of wet sacking as protection from the intense heat. They had nothing but the toweling they wore around their necks which they would hold between their teeth to prevent them inhaling fumes from the liquid steel. Both strength and precision were required to pour the liquid steel cleanly into the molds. Again, these workers were also exposed to fumes, glare and constant danger.[19]

The working conditions for making cementation steel were also intolerable, requiring workers to crawl inside the furnace to stack it for firing and again afterwards to retrieve the product. Taking out the steel was "the worst job imaginable" (it required three men working 18 hours to empty the furnace) because of the heat, the salty charcoal dust and dim light; and the steel-makers found washing away the grime impossible.[20]

As we have seen, the impetus to make steel was to produce a material that combined the hardness of cast iron with the malleability of wrought iron. The introduction of the Bessemer converter provided an efficient way of converting pig iron to steel and that steel lasted longer and

FIGURE 4.5 A puddler and his helper remove a 150-pound, near-molten ball of wrought iron from a puddling furnace, Youngstown, Ohio, 1920s. Photographer unknown. Source: https://thatdevilhistory.wordpress.com/2014/01

performed better than iron for one of the most significant applications, the making of rails. Equally, Siemens-Martin steel-making, a slower but higher volume steel-making process, made a similar contribution to the expansion of steel production. Another key factor in the progressive displacement of iron by steel was that steel-making was more amenable to mechanization. And it was via mechanization that the most immediate dangers of steel-making to health and safety were dramatically reduced. This went hand in hand with a reduction of the workforce. Henry Bessemer described the advantages of his converter thus:

> In one compact mass, we had as much metal as could be produced by two puddlers and their two assistants, working arduously for hours with an expenditure of much fuel. We had obtained a pure homogenous ten-inch ingot as the result of thirty minutes blowing, wholly unaccompanied by skilled labour or the employment of fuel.[21]

The rise of machine tools

It is into this context that we now introduce the rise of machine tools, which besides having a direct relation to the history we have just touched on—specifically in terms of the nature of work, the mechanization of labor and the human body—also reveals the complex embeddedness of knowledge and labor power in technology. The development of machine tools is also indicative of advances in metallurgy.

Machine tools (like lathes, millers, planers, borers, gear cutters) played a central part in transforming industrial work and in establishing the productive capability of modern mechanical technologies. Moreover, the enormous transformative power of machine tools, as they functioned within systems of industrial production, rapidly increased the conversion of modern materials into modern products. For most people, what machines tools are, how they work, how they have changed is of little interest and remains virtually invisible. Notwithstanding this situation, the labor power of machine tools has inscribed knowledge and force which has literally manufactured vast numbers of products of the modern world. While machine tools became crucial components in the design and operation of the production system of industrial society, they equally determined the forms, patterns, skill base and social relations of work that constituted an industrial and social ecology. But above all, the explosion of the world of goods enabled by machine tools created a quantum leap in the productive power, instrumental attainment and negative environmental impacts of industrial culture.

Machine tools were at the core of the creation of the light engineering industries—which itself was a key factor in the rise of mass production. Furthermore, machine tools, within the frame of the industries they served, placed enormous quantitative and qualitative demands on steel-makers.

The most basic machine tools were created well over 1,000 years ago—there is, for example, archaeological evidence of wood-turning lathes from the Near East of the −second century, and sword-grinding wheels have also been found from the same period.

By the Middle Ages, machine tool technology had advanced from simple pole lathes to more complex machines, like a screw cutting lathe

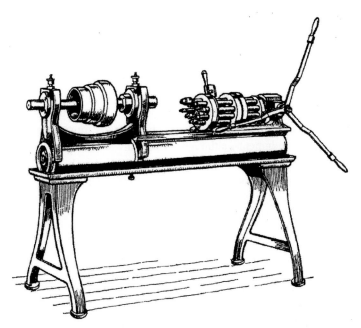

FIGURE 4.6 The Fitch turret lathe of 1848 allowed the worker to perform a variety of operations by rotating the turret which held different types of cutting tools. The turning, drilling, boring, planing and milling of steel depended on the qualities of special steels (such as tungsten steel) used to make the cutting tools. Source: L. T. C Rolt *Tools for the Job: A Short History of Machine Tools* London: Batsford and Co., 1965

(with a tool holder) used by clock-makers. By the sixteenth century, there were a significant number of machines able to work several materials across a range of mechanical functions like cutting, grinding, boring and turning. One of these machines that is frequently illustrated in histories of technology (and first appeared in Biringuccio's *Pirotechnia*) is a cannon-boring mill used in 1540.

By the seventeenth century, large and complex machines had arrived, but they were still hand- or water-powered. However, by the eighteenth century the convergence of the ability to make good cast iron, the ability to cast and bore a cylinder, and the invention of steam power changed the pattern of machine tool development dramatically. Abraham Darby's Coalbrookdale iron works was at the center of these technological advances: by 1722, cylinders were being cast for early steam engines and by 1726, a boring mill had been perfected.[22] The first modern machine shops were established by the middle of the eighteenth century.

The key to the advancement of the machine tools that had the capability of turning, drilling, boring, planing and milling steel, was the cutting tool held in the head of the machine. This in turn depended on the qualities of the steel from which these cutting tools were made.

One of the first cutting tool steels able to perform satisfactorily was Huntsman's carbon steel. Huntsman produced this in 1740 by melting blister steel to make a cast steel with uniform carbon distribution in a Chinese-like refractory clay crucible. Huntsman's innovation was as much luck as judgment—although empirically successful, he actually had no idea how important the exact percentage of carbon was to the nature and quality of the steel. The next improvement in the quality of tool steel was created by Robert Mushet in 1868, in a small iron works at Coleford in the Forest of Dean on the border between Wales and the west of England.

Mushet acquired an interest in metallurgical experimentation as a result of the tutoring of his father, who was an accomplished Scottish ironmaster. Prior to work on tool steel, Mushet had already resolved one of the main problems of Bessemer steel-making (surplus oxygen, which resulted in structural flaws in the metal), by adding Spiegeleisen (a Prussian iron ore rich in manganese) to Bessemer-made carbon steel.[23] Later, at Coleford, using a crucible, he experimented with alloying the metal of an iron from an ore rich in manganese with a powder made from ore. At that time it was called "wolfram," but later it was to be known as tungsten. When cast and forged into a bar, the resulting metal was not only a very hard steel, ideal for cutting tools, but to Mushet's great surprise was also self-sharpening. Mushet was an accomplished metallurgist; however, he was a poor businessman—his company, prophetically named the Titanic Steel Company, failed—although, with his involvement, "Mushet Special Steel" was taken up, manufactured and marketed by a Sheffield company, Samuel Osborne and Co., and became widely used throughout the world. In fact, its qualities were almost instantly recognized and universally acclaimed by the machine industry.[24]

The next advance in tool steel brings us nearer to its contemporary qualities and to an influential figure—Frederick Winslow Taylor. During the 1880s, Taylor was employed as a foreman in the machine shop of the Midvale Steel Company of Pennsylvania. Here he started investigating the metallurgy of cutting tools, their performance and relation to machine shop output. This research was given considerable support by William Sellers, President of the company, who was one of the most dominant and influential figures of the American machine tool industry.

Of particular interest to our current concern with cutting tool steel is Taylor's collaboration with a Bethlehem metallurgist, Maunsel White. At the turn of the century, together they created what became known as tungsten-chrome steel—a tool steel whose performance completely surpassed "Mushet Special Steel." The new tool "high-speed steel" produced a dramatic improvement in cutting speeds and was further improved a little later with the addition of vanadium. Modern high-speed steel had arrived and, as L. T. C. Rolt, citing an engineer writing in 1914, points out: "the new steel made every existing machine tool obsolete overnight."[25] He also pointed out that to gain the advantages of the new tool steel: "… the machines using them must be completely redesigned. They must be more robust; both feed power and drive power must be increased; hardened steel gears must replace cast gears in both drives; lubrication must be improved and finally the speed range of drives must be greater …"[26] In simplest terms, unless the machine had the weight, stability and power (which in large part came from the introduction of electric motors, which replaced steam-powered belt drives toward the end of the nineteenth century) using "high-speed steel" (which required the machines to run faster) would cause them to vibrate and then disintegrate. Thus, as a result of improvements in cutting tool steel, machine tools themselves had to be improved.

The arrival of "high-speed steel" was not the end of the story. The next development was to come from Germany in 1926, and was directly linked to the invention of tungsten carbide steel by Krupp. This was produced by the prior sintering of tungsten carbide with cobalt and then introducing this mix into the furnace with liquid steel to produce the alloy. It was put on the market in 1928 and thereafter manufactured in Germany, Britain and the USA. Its impact outdid the arrival of high-speed steel (in fact it could cut steel at a speed over five times faster than high-speed steel!). The Krupp product was amazingly hard and could only be cut with other carbide tools (which are most usually grinders). This tool steel not only transformed the speed at which metal could be cut but also cutting technique—and again this had major consequences for machine tool design (especially for feed speed and gearing). It created new performance requirements for cutting oil lubrication and for safety, as the swarf (red-hot metal cut by the tool) posed a real danger to the machinist.[27]

This account of the development of tool steel has meant that we are ahead of ourselves. We need to return to earlier histories to look in a little more detail at the history of machine tools more broadly.

The first modern machine shops were established in the middle of the eighteenth century. One figure who emerged out of this moment whose work established the standard of machine tool design and performance was Henry Maudslay. As a young engineer and a gifted designer Maudslay trained at the Woolwich Arsenal. However, he first made his mark in 1789 while working for Joseph Bramah, the inventor of the most advanced lock of its day. Maudslay fully mechanized the production of the lock, first by the design and construction of a number of special machine tools (three of which are still in the Science Museum, London) and then by the adaptation of other machines already in use. By 1797, Maudslay had set up business on his own as a machine tool maker. His first machine, now celebrated and again housed in the Science Museum, was the first in a series of screw cutting lathes. What distinguished Maudslay's work and established it as the standard, was his attention to detail and finish. He brought to the making of machine tools the level of craft skills and the precision of the scientific instrument maker. Appropriately to his objectives, he also had an obsession with accuracy (especially in terms of screw threads and the trueness of plane surfaces). The qualities Maudslay displayed and the standards he set became the foundation for the practices required to deliver all modern industrial process machine tools.

Maudslay's obsession with accuracy prompted him to make measuring instruments, like his 1805 micrometer which came to be taken as the measure of measure (and was used to settle any disputes over accuracy) and the 44 block-making machines commissioned by Portsmouth Dockyard. These, which took Maudslay six years to make, were regarded as technical wonders of their day (in fact they became a tourist attraction). They were operated at the Dockyard by a workforce of ten and produced of the order of 160,000 blocks per year. Some of these machines were still in use at the time Rolt was writing about them in 1969.[28]

Maudslay's influence extended over the entire industrializing world. As well as his machines setting the standards by which all other machines were judged, he contributed immensely to the development of engineering and industrial culture, through the men he employed and who learnt from him.

Maudslay's chief draughtsman, John Clements, set up a small engineering company of his own and became famous for the demanding construction of Charles Babbage's renowned calculating machine. He also made his lathes and a famed large planer (of 1825), which was housed in his engineering works and for a period of ten years was the largest in the world. The demand for the use of this machine was such that it created a continual source of income for Clements for the rest of his life.

Richard Roberts, employed as a turner and fitter by Maudslay in 1814, also went into business, setting up a works in Manchester as a machine tool maker. He made many machines of note, was credited as creating "more useful drilling machines" than any other engineer and was acclaimed in Manchester, which at the time was the world capital of cotton making, as the inventor of the automatic spinning mule. Later in his career, Roberts teamed up with the Sharp Brothers to form Sharp, Roberts and Co., makers of steam locomotives. This company was the first in its industry to standardize componentry. Richard Roberts is numbered among the great craft-worker inventors of the mechanical age.

Another of Maudslay's disciples of similar stature was James Nasmyth, son of a Scottish painter of note. Nasmyth was appointed the "master's" personal assistant in 1829. After Maudslay's death in 1831 Nasmyth returned to Edinburgh for a short while and then set up a business in Manchester—much in the same vein as Roberts. The business later became Nasmyth, Wilson and Co., building locomotives and making machine tools of the highest standard (a number of which were collected by the Science Museum). In 1836, Nasmyth built the well-known Bridgewater Foundry. While being best known for his steam forge hammer, which transformed the heavy forging process, his contribution to advancing machine tool design and manufacture was considerable.

Notwithstanding the place Clements, Roberts and Nasmyth occupy in the history of machine tools, and their attainments in establishing their own reputations while extending the influence of their master, there was one man who learnt his craft from Maudslay during the mid-1820s who outshone them all. This was Joseph Whitworth, who came to be regarded

FIGURE 4.7   Steam forge hammer, nineteenth century. Source: F. W. Harbord and J. W. Hall *The Metallurgy of Steel* Volume 2 London: Charles Griffin & Co., 1918

as Maudslay's equal. In common with Roberts and Nasmyth, he set up business in Manchester (in 1833). While Whitworth also contributed to machine tool design and development, it was his commitment to take Maudslay's obsession with detail to a higher level that established his reputation and enduring fame. Like Maudslay, his machine tools were guarantees of workmanship and performance. In 1856 he raised the standard of accuracy of the machine tool industry by a factor of ten (this meant taking it to 1 millionth part of an inch). In doing this, he established the standard of modern machine tool engineering—which transpired to be essential for the manufacture of sophisticated machines and mechanical technologies in the twentieth century. Of his general contribution to the engineering and the machine tool industry globally, Rolt says: "The proportions of the modern machine tool, massive, austere, strictly functional, owe more to Joseph Whitworth than any other man."[29] Whitworth also created a standard for screw threads, which rationalized thread angles. This standard ruled for well over a century and made him a household name.

Clearly, many other European nations contributed to the advancement of machine tools. However, Britain, as the first industrial nation, held a position of leadership for a considerable period. Ironically, by the time Whitworth was at the height of his fame, leadership in machine tool manufacture and light engineering was being taken over by the USA, again for reasons underpinned by historical circumstance. This history marks several generations of brilliant designer-engineers prior to the rise of "the designer" as a specific professional division of knowledge.

Looking at America

The move from colonial settlement to independence dominated the early history of colonized North America. Post-independence, there was a demand for "national development" that coincided with the rise of industrial culture in Europe, followed by its emergence in America. The political and the industrial then converged.

The impetus toward the national development of the nation generated a great deal of economic and geographic expansion. The demands of the "development" outstripped the limited labor power capability of a small workforce. The response was to design and create a technology of quality to compensate for this limitation. The result: weapons with increased firepower; agricultural machinery able to increase productivity; and system construction methods to speed the building process. This situation produced a disjuncture between material needs, market demands and the ability of industries to supply manufactured goods to meet the needs—be they for weapons, ploughs, tools or nails. The rise of a "local" iron industry was very much caught up in these events. As we shall see, its capability to respond to this situation was undercut by protectionism exercised by the British iron industry supported by the British government.

In the second decade of the eighteenth century, the American colony started exporting small amounts of iron to Britain. The reaction of the imperial power was violent and immediate—the activity of making pig, bar or any other kind of manufactured iron was banned in the American colony. This was partly because of pressure from the British ironmasters and partly because of a more general desire to keep the colony dependent upon exported goods. In contrast, filling otherwise empty ships with timber to bring back to Britain to make charcoal was regarded as something to encourage. However, the ability to enforce the prohibition on iron-making was ineffective, and by 1750 a small amount of iron was again being exported to Britain (the argument being that Britain could gain more favorable terms from America than it could from Sweden and Russia, the two main nations Britain imported iron from at the time). Change was occurring, but a number of restrictions remained; specifically, rolling mills, plating forges and steel-making furnaces could not be built. Such events of course contributed to the conditions that led to the War of Independence in 1775.

Even with both imported iron and local production, the American demand for iron could not be met. The situation was made worse by a law passed by the British government in 1785 which prohibited the export of any tools, machinery or engines, or the emigration of any skilled tradesmen, that were in any way associated with the iron industry (the latter restriction was especially unsuccessful).[30]

The combination of a shortage of skilled labor in the American economy, market demand *plus* the imperial restrictions fuelled an enormous momentum to establish an industrial culture with vibrant iron, steel and machine tool industries.[31] These demands, while functioning within restrictive conditions, were met. There were many problems on the way, but by the first quarter of the nineteenth century the goal was realized—the key to success was the use of machines in both heavy industry like iron and steel-making and light engineering.

Shortage of skilled and unskilled workers meant there was a pragmatic impetus to develop machines that could embody and extend labor power in mechanized forms. Technologically, the responses to the challenge were insightful and creative. As Brody tells in relation to the steel industry:

> Methods of smelting, refining and rolling came from Europe. But Americans mechanized the processes beyond the expectation of their inventors … By 1900, engineers had solved the major problems of mechanization: the handling of materials, integration of production stages, and continuous rolling of steel.[32]

Innovations included electrically powered travelling cranes to move raw materials to the furnace mouth and to charge it; ladle cars on rails to carry the molten iron from the blast furnace directly to the converter (which eliminated pig casting bed and its associated labor). Increasingly, iron smelting became an adjunct to steel-making as integrated works were built.[33]

Mechanization extended to the next stages of the process. From the converter, ingots were cast on rail cars to take to the rolling mill where overhead electric cranes and a mechanical plunger stripped them from the molds (an innovation introduced in the Duquesne Works

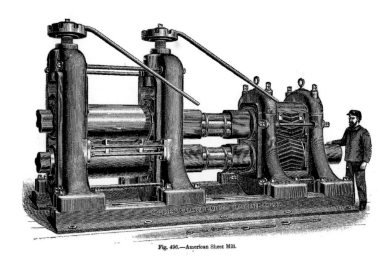

Fig. 496.—American Sheet Mill.

FIGURE 4.8 Sheet rolling mill, late nineteenth century. Source: F. W. Harbord and J. W. Hall *The Metallurgy of Steel* Volume 2 London: Charles Griffin & Co., 1918

in Pennsylvania in 1889, which were purchased by Andrew Carnegie and subsequently introduced into all his mills). In the rolling mills, automation became the norm for depositing the ingots onto the rolls, for rolling, reheating, stamping and cutting. Processes that were previously separate became fully integrated, so that for example, steel rods could be rolled directly from billets, "until it is a familiar sight to see a billet, one end still in the furnace – its length in all the reducing passes of the mill, and the other end coiled on the reel, a finished wire rod," so commented F. H. Daniels in 1891, one of the inventors of this process.[34]

By the 1960s casting was further mechanized with the introduction of the continuous casting process in which the molten steel is directly cast into semi-finished steel shapes, thus cutting down on handling and eliminating the need to reheat ingots. The mechanization of the rolling of sheet steel and tin plate was not as rapid as it was for rod and bar, although by 1899 travelling cranes and electric power were also in use in this sector and machines had been developed for annealing, pickling and polishing.[35]

An entirely new type of light engineering industry was created out of the same circumstances. It took the ideas, methods and standards of

FIGURE 4.9 Travelling ingot charger, late nineteenth century. Source: F. W. Harbord and J. W. Hall *The Metallurgy of Steel* Volume 2 London: Charles Griffin & Co., 1918

people of the likes of Maudslay and Brunel, as well as French and German gunsmiths, and learnt from them. In so doing, the standards of the British fitter engineers were adapted and systematized, resulting in viable methods of standardized manufacture wherein all components made were completely interchangeable. This meant that a small semi-skilled workforce could assemble large batches of the same product quickly and without the need to adjust any of the components. Obviously interchangeability was one of the important precursors of mass production. By the mid-nineteenth century, these methods were becoming known internationally as the "American System of Manufacture."

The light engineering industry, centered in New England, initially focused on small arms production—the rate-of-fire capability of weapons was again a design response to the "need" to give a small number of soldiers a large amount of firepower. A whole range of technologies, products and industries emerged from the skills and machine capability of this industry. Equally, a number of iconic figures emerged, like the gunsmith engineers Eli Whitney, Samuel Colt, Frances Pratt (who was to join forces with Whitney to create the renowned Pratt and Whitney Company) and the famed mechanic, Elisha Root.

Such was the quality of the engineering in the arms industry during this period that the machine tool maker Robbins and Lawrence, formed in 1838, were by the early 1850s exporting machines to the Enfield Armoury in England. Another machine tool maker of great repute, established in the same period (1833), was Brown and Sharp. The company first became internationally known for their manufacture and export of rules, calipers, vernier gauges and other measuring instruments. This company broadened the marketing of machine tools with a representational regime in which machine design and its qualities were only part of the communication. The company realized, as Tim Putnam points out, that in the first instance it was selling a corporation, a reputation and an image that all centered on the qualities associated with the product.[36]

The next figure to acknowledge is William Sellers, who will return us directly to the steel industry.

Sellers was at the center of the rise of the machine tool industry in Philadelphia, a city that established a powerful location of engineering outside of New England in the 1850s and 60s. Leading the way for these developments, he set up a machine shop in 1848. According to Rolt, his reputation matched Whitworth's and he had the same kind of impact on his industry.[37] His machines were of the first order, they won prizes and, like Whitworth, he created standards adopted by government. It is not, however, in this role that he contributes to our account, rather it is in the position he gained in 1873, that of President of the Midvale Steel Company. More specifically, it was Sellers' hiring of F. W. Taylor that was to have significant consequences.

Taylor had been attracted by the reputation of Sellers. In 1878, after finishing his apprenticeship as a pattern maker and machinist in a small machine shop in Philadelphia, he joined Midvale as a journeyman pattern maker, although initially working as a laborer. In eight years, Taylor progressed from this position to chief engineer of the works, while also gaining a Masters of Engineering via evening study at the Stevens Institute.[38] Taylor actually brings together the three elements of our account: steel, machine tools and labor. This connection was as a result of his applied research, first supported by Sellers and later by the Bethlehem Steel Company. His research spanned the design of work (especially the organization of labor processes), management and materials. Extended over a period of 26 years, it had a global impact on industrial culture, industrial relations and workplace ecology. The ghost of Taylor still walks and his afterlife is protracted.

Machine work

Taylor's work on tool steel directly informed his influential publication, *On the Art of Cutting Metals* (1906). This work had begun in 1880 at Midvale with a study of the operation of a 66-inch diameter vertical boring mill. Out of this initial research, a whole field of experimentation was created over an extensive period. Taylor quantified this in 1911 as between 30,000 and 50,000 experiments, in which he used about 800,000 pounds of steel. Two questions drove the entire exercise: "at what cutting speed shall I run my machine?" and, "what feed shall I use?" What this really added up to was the ambition to displace the judgment of the machinist with science and rules. Taylor explained that the "complexity of the research" was why it took so long—this complexity was empirical and centered on the fact that there were 12 variables.[39]

At the completion of this research, Taylor ended up with a special machine shop slide rule that enabled the machinist to do a very quick calculation in order to set up his/her machine. Additionally, because the machine tool makers had not designed and set up their machines with a full knowledge of what the optimum cutting speeds were to be for specific tasks, there was also a task of resetting the machine—this with modified pulleys, etc., so it would run at the required speed.[40] All of this work was done in order to speed up machine output (which included the elimination of tool remaking and resetting) and eliminate the "rule of thumb action" by the machinist (which completely undervalued the extensive tacit knowledge of skilled workers).

Research was, of course, only part of Taylor's project. He also strove to replace the judgment of workers with "science" in all possible contexts, as well as using "scientific management" to select the most appropriate worker for the task—in this respect he had an ideological mission to restructure work so that the worker was placed in service to, and directed by, the machine. In other words, the worker was totally disempowered and completely instrumentalized.[41]

To illustrate his thinking on work methods, Taylor frequently cited the case of the pig iron handler "Schmidt." Taylor claims Schmidt as a scientifically selected worker, chosen on the basis of his history of compliance in the workplace and his strong desire for more money. Of him, he said: "… the pig-iron handler is not an extraordinary man difficult to find, he is merely a man of more or less the type of the ox, heavy both mentally and physically."[42] The objective of scientifically managing Schmidt was to increase the volume of pig iron he loaded onto a rail car from 12.5 tons to 47. This was done with the power of a bribe—Schmidt was offered a higher rate than other workers in return for complete compliance to the instructions of "a man standing over him with a stop watch." Of science and this method, Taylor said: "the man suited to handle pig iron cannot possibly understand it, nor even work in accordance to the laws of science, without the help of those who are over him."[43] We should remind ourselves at this point that Taylor was the most influential theorist of workforce management there has ever been. His agency has rested not so much on the practicality of his methods but more on their "ideological" mobilization and adoption as "common sense."

Fundamentally, Taylor set out to reduce the worker to the status of an interchangeable machine component, fully compliant with managerial instruction. His aim was to eliminate "soldiering"—by which he meant any form of activity whereby the worker strove to retain some control over the job or any action that resisted the will or instruction of management. Needless to say, as far as socialist theorists of labor were concerned he was a pariah.

What Taylor failed to recognize or value in his research was the experience and the knowledge of the worker. As Gordon writes:

> Artisans at ironworks learned their metallurgy by doing rather than by reading or talking. Although such tasks as stacking pigs or wheeling coal call for a strong back and little thought, most ironworkers had to exercise vigilance, judgement, and metallurgical knowledge based on judgement. Some tasks that look like routine labour actually involved particular skills. Artisans had to know the proper amount of moister to put on the sand that received the molten metal from the blast furnace or, when working on its charging platform, how to distribute ore and fuel to keep the chemical reactions within the furnace going smoothly. Artisans' errors could ruin the metal, damage the furnace, endanger the furnace crew. In addition to their heavy physical labour, finers, puddlers had to control invisible chemical transformations without the aid of analysis of the materials they used or of instruments to show conditions within their furnaces.[44]

The failure to recognize the tacit knowledge of workers has actually been one of the fundamental flaws of the modern industrial labor process. It was not simply manifest at the early stages of the second industrial revolution, as illustrated by the work of Mike Cooley and David Noble in the 1970s and 1980s on the deskilling of fitters, machinists and turners by the computerization of machines.[45] The failure of managers to acknowledge workers' tacit knowledge has had profound directional consequences on both workplace culture, the self-image of workers and the products of their labor. On the one hand, it points to the lost potential of a might-have-been-otherwise industrial history, rather than one which was based on de-powering workers on the basis of mistrust, the fear of their organized and individual resistance and the desire for total control. Although in recent decades more "progressive and enlightened" employers have recognized the contributions that direct experience on the shop floor can bring to improving productivity and product quality, this has been in an epoch when the widespread culture of skilled workers had been broken. Certainly, the contemporary valuing of this knowledge has not been culturally enabling and formative of new cultures, rather it has been an appropriation that economically was a final act of theft (if not always felt so inter-personally).

More generally, Taylor was a dangerous, obsessive and completely uncritical servant of capital logic, who in many ways was duped by the people he served. Time has passed and his manic relation to efficiency and militaristically designed disciplinary regimes (this to drill the body of the worker into machine compliance) now increasingly appears as absolutely unsustainable social and ecological practice.[46]

Superficially, the speed of cutting metal, the speed of machine operation and the speed of the laboring worker were all actions in the total service of speeding up steel production, the volume used and the speed of the growth of industrial capitalism. This "commitment" is fundamentally why Taylor gained so much support from Sellers at Midvale and later, the Bethlehem Steel Company. However, without knowing it, what he was doing more fundamentally was speeding up the rate of transfer of metaphysics from the human mind to the machine—a project that the machine tool industry was always, and still is, implicated in (be it perhaps unwittingly, or certainly with an extremely limited grasp of the implications of its actions beyond the limited horizon of machine and factory performance). No machine tool designer, to our knowledge, has grasped that they were in fact intervening in and transforming the nature of an ecology of mind with profound consequences for the future of the human race. Echoing earlier remarks, science and technology are still claimed by scientists and technologists, at least publicly, as being under human control, whereas research from many other disciplines is showing that these domains are taking on a life of their own, to which human beings are simply reacting.[47] In the end, while material impacts are significant, this is what is most unsustainable of all.

Conclusion: Uncontrolled control

Machine tools and industrial machinery in general, have been subordinate to a capitalist logic that sought to create, and often impose, systems unable to be disrupted by worker vices, attitudes, dispositions or resistance. The machine was viewed through a utopian lens. The coming of mechanization offered the prospect of uninterrupted work around the clock, without complaint, a need for food, rest or time to tend the needs of nature. This utopia travelled through time: it was stamped on the waterwheel, on Joseph-Marie Jacquard's automatic loom of 1804; it was the dream of F. W. Taylor's entire project and Henry Ford's assembly line; it was the inspiration for the rise of paper tape machine tool operating programs that led to contemporary computer-driven numerically controlled machine tools, and of the rise of CAD (computer aided design) and CAM (computer aided management). It was also part of the vision of FMS (flexible manufacturing systems) linked to fully automated factories, computerized sales feedback, stock control and even market data programming product form.

This utopia has in fact not delivered the "leisure society," or any other idealized future projected in the not-so-distant past. Rather, it has been at the core of anthropocentric unsustainable practices of production whereby excesses of the world of goods and cultures of unrestrained material desires drive material impacts (emissions, waste, pollution and the like) that negated futures. Making has been an unmaking, a driver of ecological and environmental destruction, and the means by which futures have sought to be made that have de facto defutured. Yet this history is also replete with lessons to learn which are able to fuel actions, knowledge and skills toward recovery.

One writes and thinks of this kind of history with mixed feelings— of pride, pleasure and sorrow. It marks a certain pinnacle of human achievement in terms of the exercise of the craft of the engineer. Imagination, a facility of the hand and eye, the ability to use and dominate the machine—all combined to create objects of great elegance and functional capability that are celebrated in histories of engineering, technology and design, as well as in museums of science and industry. One can gain pleasure from standing before such machines and wondering at the form, ingenuity and skill of the maker. Yet the obsession that brought the object into being was blind. It saw nothing but its task, and here is the sorrow. The machine was designed against its user in a passage of developments which aimed to negate the exercise of knowledge and intervention in the labor process. The development of the highly refined machine skills of the worker was but a passing moment. The fact that knowledge was being transferred from mind to machines and to machine-functions was not a concern. In looking at the kind of machine tools cited here, we are viewing objects of a vanished culture. They might as well be thousands of years old, which is not to say there is nothing to discover or recover. The sadness then is for the "might-have-been," the "what-is-lost" and the myopia of the technically perceptive.

Notes

1   See Walter Mignolo *The Darker Side of Western Modernity* Durham, NC: Duke University Press, 2011.

2   This is specifically because it is unable to smelt iron, and so is dependent upon scrap or directly reduced iron (DRI) (pig iron substitute) and while not as energy-hungry as steel produced in an integrated steel works, it is still energy-intensive. Heat is created in the EAF by the resistance of metal to the current induced from carbon electrodes immersed in the charge. No other fuel material is added. It is a highly controllable process, and one where reduction, oxidization, slag and temperatures are all constantly monitored and managed.

3   Leslie Aitchison *A History of Metals*, Volume 2, London: MacDonald and Evans, 1960, pp. 340–3.

4   Howard G. Roepke *Movements of the British Iron and Steel Industry 1720 to 1951* Urbana, IL: University of Urbana Press, 1956, p. 21.

5   Harry Scrivenor *History of the Iron Trade* London: Longman, Brown, Green and Longmans, 1854, p. 183.

6   The modern Basic Oxygen Steel-making (BOS) takes the converter process and replaces atmospheric cold air with a lance, introduced from the top (or in now more advanced converters from the bottom), which injects almost pure oxygen (99.5 percent) at supersonic speed into the molten metal—this results in high temperatures and rapid oxidization to reduce the absorption of nitrogen. Iron is converted to steel in less than one hour.

7   The modern coke-burning blast furnace has largely replaced the open-hearth process. Those still operative are not only technologically outdated but are now regarded as environmentally undesirable. The historical developments of the technologies sketched above paid no heed to harmful emissions, contaminated liquid or solid waste, workplace safety, immediate health hazards or any other environmental factors.

8   Ure *Dictionary of Arts, Manufacturing and Mines*, pp. 1108–11.

9   The more carbon that was expelled the higher the metal's melting point. This process produced a very high iron loss.

10 Eric Hobsbawm, for instance, writes on the "immense accumulated historical advantages in the underdeveloped world, as the greatest commercial power, and as the greatest source of international loan capital; and had, in reserve, the exploitation of the 'natural protection' of home market and if need be the 'artificial protection' of political control over a large empire." A little later he says: "Iron and steel relied on the Empire and the underdeveloped world, like cotton: by 1913 Argentina and India alone bought more British iron and steel exports than the whole of Europe, Australia alone more than twice as much as the USA". *Industry and Empire* Harmondsworth: Penguin Books, 1969, p. 191.

11 Antonio Gramsci "The Study of Philosophy" in *Selections from Prison Notebooks* (trans. and edited by Quentin Hoare and Geoffrey Nowell Smith) London: Lawrence and Wishart, 1978, pp. 321–77.

12 Donald B. Warner *The Traditional Chinese Iron Industry and its Fate* Copenhagen: Nordic Institute of Asian studies, 1997 (web version), p. 7/60.

13 Ibid., p. 8/60.

14 Vannoccio Biringuccio *Pirotechnia* Cambridge, MA: The MIT Press, 1958, p. 373. On the sensory intelligence of the primitive smith we note Needham's remarks: "The primitive smith added or removed carbon, as well as other elements, knowing nothing of the chemistry of what he was doing, but he knew what he wanted to obtain, and had rough tests as well as skilled and experienced eye." Joseph Needham *The Development of Iron and Steel Technology in China* London: Newcomen Society, 1958, p. 11.

15 Robert B. Gordon *American Iron* Baltimore, MD: The Johns Hopkins Press, 1996, caption 10–2, p. 224.

16 Puddling description from Gordon op. cit., pp. 140–7 and Norman J. G. Pounds *The Geography of Iron and Steel* London: Hutchinson and Co., 1959, pp. 19–20.

17 Pounds op. cit., p. 20.

18 F. W. Harbord and J. W. Hall *The Metallurgy of Steel* Volume 2, London: Charles Griffin and Co., 1918, p. 526.

19 Pounds op. cit., pp. 181–4.

20 Pounds op. cit., p. 176.

21 Pounds op. cit., p. 22.

22 See L. T. C Rolt *Tools for the Job: A Short History of Machine Tools* London: Batsford and Co., 1965.

23 The manganese attracted the oxygen and held it in the slag as manganese oxide. L. T. C Rolt *Victorian Engineering* Harmondsworth: Penguin Books, 1970, pp. 182–3. Two observations on Bessemer are worth making. First, while Bessemer was a gifted inventor, he actually only had a very limited amount of metallurgical knowledge, and second, he never gave Mushet proper acknowledgement or financial rewards for his work.

24 This story of the production of this steel, as Rolt tells it, still a "secret to this day." Ibid.

25 Ibid., p. 200.

26 Ibid., pp. 200–1.

27 Ibid., p. 227.

28 Ibid., pp. 200–1.

29 Ibid., p. 119.

30 See Gordon op. cit. for a more detailed account of these events.

31 Rolt *Tools for the Job*, p. 138.

32 David Brody *Steelworkers in America: The Nonunion Era* Cambridge, MA: Harvard University Press, 1960, p. 11.

33 Ibid., pp. 9–14.

34 Ibid., p. 12.

35 Ibid., pp. 9–14.

36 Tim Putnam "The Theory of Machine Design in the Second Industrial Age" *Journal of Design History* Volume 1, No. 1, 1988, pp. 32–3.

37 Rolt *Tools for the Job*, p. 175.

38 Harold Parsons, Foreword, Frederick Winslow Taylor *Scientific Management* New York: Harper and Row, 1947, p. ix.

39 This account is drawn from Taylor's *Scientific Management*. His 12 variables were: (i) the quality of the metal, (ii) the chemical composition of the steel, (iii) the thickness of the shaving cut, (iv) the contour edge of the cutting tool, (v) the volume of cooling fluid used, (vi) the depth of the cut, (vii) the duration of the cut, (viii) the lip/clearance angles of the cutting tool, (ix) the impact on the work of machine chatter, (x) the diameter of the casting or forging being cut, (xi) the pressure of the chip/shaving on the cutting tool, and (xii) the pulling power and speed and feed changes of the machine tool. Ibid., pp. 107–8.

40 Another associated research project of Taylor's was a nine-year study of machine belt drive tension—Shop Management p. 125 of *Scientific Management*.

41 Taylor's methods were, in part, influenced by those of his contemporary, Frank B. Gilbreth, a production engineer who studied human movement in great detail, in order to apply the knowledge to the scientific management of work and the worker.

42 Taylor op. cit., p. 137.

43 Ibid., p. 48.

44 Gordon op. cit., pp. 18–19.

45 Mike Cooley *Architect or Bee? The Human Technology Relationship* Slough: Hand and Brain Publications, 1980 and David F. Nobel *Forces of Production* Oxford: Oxford University Press, 1986.

46 It is no mere coincidence that an analysis of the Taylor Society of 1918 disclosed that a third of the membership was employed the US Ordnance Department or that the US Army's approach to the integration of military and industrial management in both the First and the Second World Wars was heavily influenced by his theories. See Merritt Roe Smith "Army Ordnance and the 'American System' of Manufacturing, 1851–1861" in Merritt Roe Smith (ed.) *Military Enterprise and Technical Change* Cambridge, MA: MIT Press, 1987, pp. 14–15.

47 See Tony Fry "The Autonomic Technocentricity of Computers" in *A New Design Philosophy: An Introduction to Defuturing* Sydney: UNSW Press, 1999.

FIGURE 5.1   Crane hook for lifting locomotives, assembly plant, Indiana 1919. Photographer unknown. Source: Martin Grief, The Austin Company, 1978

5 FULLY MECHANIZED MODERNITY

n the previous chapter, we looked at how improvements in the capabilities of steel contributed to diversification and expansion of the machine tool industry, which in turn was a major driver of industrial modernization. We also saw how complex were the relations between steel as a material, its applications, the labor processes associated with it, and the "ecology of mind" that provided the spark for the cumulative innovations that are gathered under the banner of "the modern."

This chapter stays within the same arena of concerns but moves on to consider some of the more overt world-forming and transforming products of steel. Specifically, we will look at how steel, in some of its most significant applications, has been a key force in the making of the modern world. The applications are: war machines, railways, motor transport, road building (as concrete reinforcement), construction (as building framing, cladding and roofing), and material infrastructure (bridges, pipes, pylons, etc.).

Obviously, every one of these applications has a substantial history in its own right. Our aim is not to summarize these, but to draw out some of the trends that contributed to steel becoming a taken-for-granted material of modern life. In so doing our motive is to denaturalize steel's uses, which means focusing attention on much that has become so familiar as to become invisible, in order to expose its uses to new questioning.

As we have seen, global production of steel is constantly increasingly and its material forms proliferate. Likewise the industry is in constant flux internationally. "Steel" seen in this context does not simply refer to an inert material, but also to the interests of an industry, its linked financial institutions and the material's users. The impacts of steel flow from the totality of the environments that produce it and the environments it contributes to producing.

Holding steel to account, within this expanded definition, indicates at best an ambiguous record. Yes, steel did constitute futures; it did make things possible that previously were not. But equally, it became an agent that took futures away, and as such, was a force that defutured. It would be unfair to backload present-day knowledge and the wisdom of hindsight onto past intent and interests. However, during periods of rapid industrial expansion, iron and steel-makers did fail to pay due regard to the discernible damage they were doing to the physical environment, to the immediate dangers of industrial processes in the workplace or to the cumulative damage to the health of workers. While in contemporary advanced industrial cultures, a crass disregard for environmental impacts and workers' health cannot be got away with, the disposition towards absolutely privileging wealth-creation remains dominant. This results in taking environmental action that strives to sustain the economic status quo. Effectively what dominates in this arena is "sustaining the unsustainable." Thus while environmental actions taken are worth taking, they are conservative and do not lead to fundamental change.

Contrary to the inscribed trends indicated, steel-makers and users need to develop a sensibility willing to shift into another register—effectively this means negotiating the divisions between creation and destruction. Making critical judgments on options for radical change will become essential in emergent global environmental circumstances. Corporate interests, major perceptual reconstruction and the interests of environmental management will all converge as the imperative to gain the ability to sustain becomes more urgent. What this means is not simply a more "eco-efficient" and lower impact steel industry, still locked into a developmental logic that, at best, just slows the rate of environmental damage. Rather, it means finding ways for steel to make a substantial contribution to advancing futuring processes. For this to happen the mind-set, practices and products all have to change (these issues will be further explored in the final chapter).

From the perspective of the criticality of sustainment, the central question will become: "why use steel?" All other question, like: "how to recover it for reuse or recycling" can only follow on from a fundamental questioning that treats the issue of "why" rigorously.

Issues of use may seem obvious enough, but in fact demand considerable thought. Use is not merely a mundane or practical question; rather it comes from "the world of uses." We are cast, by upbringing, education and training, into the ways of operation of the functional world. We use the things of this world and they use us.

Although this statement sounds somewhat philosophical, it is unavoidable.

What we have to come to terms with is this: that "the way the world is" and our "induction into it" circumscribes (and thus limits) what we take to be freely chosen actions. We have used the term "common sense" to identify the embeddedness of the force of ideology. As soon as someone says: "it's common sense to use steel for this or that use" they are extending the world of their induction. In this context, "fundamental questioning" means questioning what is taken to be the accepted as "common sense" in use. In turn, this means denaturalizing taken-for-granted ecologies and relations/worlds/material actions and connections of uses. Put more concretely, this means asking far more critical questions of steel's appropriateness to particular applications by giving more weight to the immediate and long-term environmental impacts of material extraction, iron and steel production and use.

Now to a brief history of some of the dominant uses of steel within the qualifications made.

At the beginning: Weapons, war, ships and bodies

Bronze weapons and tools prefigured the first products of iron. The limits of bronze fed the desire among armorers for a harder metal, one able to hold an edge longer.[1]

Between −900 and −612, the Assyrian "iron army" created an empire by military domination of the ancient Near East. Their iron armor provided greater protection, while iron weapons had a superior cutting edge and greater penetrating power, be they axes, arrows, spear heads, lances, swords or daggers. The strategic advantages of iron made this the largest, best equipped and most well-organized fighting force that the world had encountered.[2] The Assyrian "iron army" thus became the model for all the military powers of the period, including the Greeks. At the end of this era another kind of major military force arrived—the first large-scale navy. This was created by Persia, which assembled a fleet of between 400 to 800 ships.[3] While pre-dated by the Greeks, the Persian navy was larger, better equipped, more skillful and far more dangerous. The Greeks responded to the challenge and an uninterrupted pattern of competition for naval power was set. The eventual victory of the Greeks over the Persians had to await the coming of Alexander.

Military history tells us that after the accession of Alexander the Great in 404 there was a military revolution in which the "Greeks changed the nature of ancient warfare and produced one of the finest armies in the military history of the western world."[4] Alexander took the art of war to levels that would rarely be excelled for more than 2,000 years.[5] His tactical device of the "hammer and the anvil" remains a basic military ploy even today. The form of discipline he installed, the quality of weapons used, the speed of movement of his armies to the battle site and in battle, the mode of streamlined logistical support he devised, and the significance given to planning and intelligence—all of this set the standard.

The influence of the Greeks upon Roman military culture is evident.[6] What is of particular interest to us is the technological support that underpinned the Roman military machine. This was most visible in the Romans' development of the *fabrica*—a name applied to both "the great military forge" and the "college of armourers."

At the start of the second century, the smiths who travelled with the Roman Legions were organized into companies headed by the *primicerius*. These companies made arms for the legions in workshops and forges, called *fabricae*, which were erected at every major camp and garrison. They were also established in larger organizational forms to serve the arms needs of those legions who were the controlling imperial power for entire provinces. The scale of this industrial activity contributed an enormous amount to the development of the infrastructure, economy and cultural life of the cities where they were established.

All the weapons produced at the *fabricae* were handed over to an official who stored them in an arsenal and controlled their release. Here is Scrivenor on the topic:

> ... to prevent any abuse in this important branch of the military economy, and to insure proper and methodical management, no person was permitted to forge arms for the imperial service, unless he were previously admitted as a member of the society of the Fabri; that, to secure the continuance of their labours after they had been instructed in the art, a certain stipend was settled on each armourer, who (as well

as his offspring) was prohibited from leaving the employ till he had attained the office of *primicerius*; and, finally, that no one might quit his business without detection, a mark or stigma was impressed upon the arm of each as soon as he became a member of the *fabrica*.[7]

By degree, an arms industry became a structural feature of the economy and culture of all Western nations. The expansion of Western powers played a large part in globalizing this mostly metal-based industry. One particular event changed the direction, social and environmental impact of this industry—the introduction of the Chinese invention, gunpowder, to Europe at the end of the thirteenth century.

Gunpowder gave a great deal of momentum to arms production, the iron industry, metallurgy and the technologies of working and machining iron. Correspondingly, it was also to significantly increase the demand for charcoal production because the production of iron increased and because charcoal was one of the primary ingredients of gunpowder.

The introduction of cannon into widespread military use initiated a pattern of investment and technological development that still continues—barrel weight, metallurgy, casting, machining and the lifespan of weapons, as linked to the their range and rate of fire, all became enduring concerns. The same is true of projectile size, speed, cost, weight and impact performance. Thus from the cannon a whole range of artillery pieces developed, including: the mortar and howitzer (both of which marked the arrival of a capability to hit a target, not in a line-of-sight, via a calculated trajectory); field guns; heavy, siege, mountain, naval and coastal artillery; and then more latterly, mechanically propelled guns, tanks, aircraft cannon plus anti-aircraft and anti-tank guns. There is obviously also a parallel history of muskets, pistols, carbines, rifles, machine guns, sub-machine guns.

The designing consequences of heavy weapons and small arms were massive. Besides creating the ability to kill large numbers of people, they totally reconfigured the space between combatants and non-combatants. They also transformed the nature of the war machine and conduct of war in general. Not least among these changes was the combination of new weapons within disciplinary formations that systematized their use. Cannon, for instance, were organized into batteries in order to lay down a pattern of fire. Likewise, men were initially organized into squads so that the field and rate of fire could be fully coordinated. Rapid fire small arms, and especially the machine gun, of course, made this form of organization redundant.

What has slipped from view is the fact that it took considerable time for cannon to be manufactured to military standard of production and regimented forms of use. Early cannon were not simply arms among arms, but the special province of craftsmen, who were extremely protective of their secret knowledge. As Robert O'Connell, citing A. R. Hall, points out:

> ... those who cast the cannon would also serve them in battle, procuring powder and shot as well ... They were not soldiers but more akin to alchemists, members of a guild which, disdaining military discipline, passed its secrets to apprentices only, jealously and under oath.[8]

These men, it should be remembered, were among the most technically advanced craftsmen of their day. Their industry also became a significant part of the economy. There were, for instance, ten gun-casting foundries in Sussex, England in the mid-sixteenth century producing 500 to 600 tons of steel per year. Within 50 years, this output had doubled.

At the time of their introduction cannon actually had as much psychological as physical impact. The violence coming from a disembodied source, the noise of cannon firing individually "at will" or as a barrage, the sound of a projectile travelling through the air and the thud as it hit ground (and then later the blast of exploding shells), the scream of shattered bodies or the indescribable noises of dismembered horses, the smoke, the smell. All of this fundamentally changed the character of battle. Skill at arms, courage in the face of the enemy, valor on the field of battle, the honor of combat, all such notions lingered, but basically became meaningless once war became conducted with weapons of great force that dealt out death at a distance, technologically and by calculation. It was not really until the First World War, with its mass-produced death by the machine gun, that the fictions of the conduct and values of war were finally erased from the imaginations of the public, politicians and the military high command.

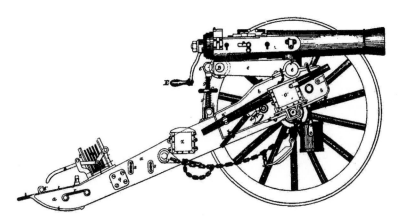

FIGURE 5.2 Gatling gun, forerunner of the machine gun. Source: John Ellis *The Social History of the Machine Gun* Baltimore: The Johns Hopkins University Press, 1975

FIGURE 5.3 Nordenfeldt gun, similar to the Gatling gun. Source: John Ellis *The Social History of the Machine Gun* Baltimore: The Johns Hopkins University Press, 1975

FIGURE 5.4 A Corporal of the British Machine Gun Corps at a machine gun post in a captured trench at Feuchy during the Battle of Arras, April 1917. Photo by Lieutenant John Warwick Brooke. Reproduced with permission of Imperial War Museum © IWM (Q 5159)

FIGURE 5.5 One of the first tanks to go into action, Battle of Flers–Courcelette, September 15, 1916. Photo by Lieutenant John Warwick Brooke. Reproduced with permission of Imperial War Museum © IWM (Q 5575)

Evidence to the contrary had been available for a long time. The American Civil War, the Crimean War and the Boer War all delivered illustrations of horrors to come; however, and without question, the one most brutal use of the gun (in its myriad forms) was its widespread use against native peoples armed only with spears or bows and arrows. This genocidal history is etched in the past and in the memory of every colonized nation on the face of the planet. For over 500 years, the gun was deployed against people who were in their rightful place. One cannot begin to imagine how different our planet would have been without this bloodbath, occupation, conquest and destruction. The damage done to the social ecology of the human species is such that it will travel with us until the end of time. The critical history of colonialism tells us that Eurocentric self-deception produced a condition whereby genocide on a vast scale went "unseen" for centuries. One could even say that one of the fundamental traits of "humanity" has been its ability turn its back on what it does not want to see and to live in "blissful unawareness."

Such remarks seems to stray—what have they to do with steel? We can see its presence and agency at every junction, we can tell ourselves that an inert material is blameless, yet it is not simply one material among many. Uniquely, and for eons, iron and steel have been present at every act of human-induced planetary destruction, transformation and creation. Wherever, whatever or whoever "man" strikes, for good or ill, iron or steel is to hand; it fills the space between the exercise of will and the point of impact.

Technology of modern weapons

The development of steel, the rise of modern machine tools and the ability to produce components *en masse* changed the technology and efficacy of guns. Rifling is an example—it was introduced for the barrels of both cannon and small arms, and it delivered great advances in weapon accuracy. Additionally, the combination of structurally stronger barrels and breaches together with advances in explosives meant that the size and force of explosive charges for shells and bullets could be increased. This in turn increased muzzle velocity and projectile range. One of the earliest and most dramatic examples of this came from the company that was to become Europe's leading armorer: Krupp.

The Krupp 6.5-centimeter crucible steel cannon, first made in 1847, after a slow uptake, became a key weapon of the German war machine and the rise of power of the German state.[9] The introduction and development of the machine gun in the mid-nineteenth century equally depended on the quality of steel—not least because although rapid fire generated a great deal of heat, the barrel still had to remain true and its bore size stable.

At the same time as the accuracy, firepower, range and destructive capability of battlefield weapons were improving, there were major developments in naval armaments and ships.

In 1861, in the course of the American Civil War, both sides introduced iron-clad warships: the Swedish inventor John Ericsson designed the single-turreted USS *Monitor* as well as the central battery Confederate ship, the CSS *Virginia*. These ships, combined with the arrival of exploding shells (first used in 1853), heralded the death of wooden warships. After the destruction wrought by these ships on their wooden counterparts, there was a significant shift in US defense policy and in the US arms industry in the 1870s and 1880s. The "intense interest" expressed by the US Naval Ordnance Bureau in all-steel breech-loading guns (like Krupp's) and in armored warships played a significant part in prompting both the Midvale and Bethlehem Steel companies to adopt open-hearth methods of steel-making so as to increase output. At the same time, government contracts stipulated that modern steel-making techniques were required for a contract to be awarded.[10] More generally, as David Lyon observes, the demand for ships as weapon platforms, long range guns, speed and armor "had a profound effect upon metallurgical techniques."[11] He also points out that the demands of the world's navies led to "the creation of bigger foundries and steel mills."[12] To this must be added other developments—most significantly, the demise of the sailing ship and the ascent of steam power.

It took from the opening of the nineteenth century until 1837 for the steamship to become fully viable as an ocean-going vessel (in that year the 600hp paddle-wheeled steamship *Sirius* crossed the Atlantic). By the 1850s the age of steam at sea had arrived—the icon of this moment being the 11,500hp *Great Eastern*, the last project of that great Victorian engineer, Isambard Kingdom Brunel. A few years earlier, in 1843, Ericsson, as we have already noted, designed the USS *Monitor*—the first propeller-driven warship (a technical development that was essential for the rise of all large, armored and large-gun warships to come). Not long after steam came the introduction of electrical power—which was crucial for the development of the submarine.[13] Four other innovations of the period should also be registered: the breech-loading gun (which in the case of the Royal Navy was in 1873);[14] the introduction of floating and submerged mines; the introduction of torpedoes; and advances in explosives technology.

In Europe, the commitment to a steel navy went even further than in America. The British Naval Defence Act of 1900 specified that the Royal Navy had to be equal to the combined strength of any other two navies (in the 50 years prior to this Britain had become the world's largest merchant and warship shipbuilder). Half the warships in the world were British built, and they were constructed at twice the speed of any other nation's shipbuilding industry.

The Defence Act had two consequences. First, it precipitated the introduction of the most powerful machine of war that had ever been created—the Dreadnought, the first of which was launched in 1906. These warships were markedly larger, faster, more heavily armored and better armed than any other ship afloat.[15] They effectively made every other warship in the world redundant, including the other 40 ships of the British Navy. It was claimed that the coming of the Dreadnought struck fear and panic into the hearts of the commanders of every navy on earth. The Dreadnoughts signaled the arrival of the modern battleship, a type of ship and concept of naval warfare that was to dominate for 50 years.[16] They also prompted a very aggressive arms race between Britain, France, Germany, Japan and the USA. As is always the case, this was also a race of materials development, machine tool advancement and industrial system expansion. It is thus no mere coincidence that in this period world iron and steel production increased by leaps and bounds.

The rush to the First World War was on!

The second consequence of this moment was the ascent of big capitalism whose economic objectives transcended national interests: the behavior of the arms and steel industry was a significant instance of this. The best-known example is Krupp, who supplied steel armor plating to almost every navy on earth. Krupp's steel considerably reduced the thickness and weight of armor—from 60-centimeter hardened wrought iron to less than 15-centimeter thick steel.[17] Obviously these reductions markedly increased the speed of warships (which prior to this still remained slow). The counter-move was the ascendance of the torpedo-armed submarine, which came into its own in the First World War, and the introduction of the motor torpedo boat, which had a slightly longer history.[18] The torpedo was responsible for significant design changes of warships, including where armor was placed, the creation of internal sections and the subdivision of sealed compartments.

Modern steel-based weapon systems, mines, grenades and projectiles cannot be understood simply in terms of the history of military technology. They have also had a profound impact as drivers of the expansion of the industry (as we have seen), on the global physical environment, on the social ecology and mind of humankind.

Nowhere was the visibility of the environmental destruction more graphic than in the First World War. The major battlefields like Verdun, the Somme, Arras, Ypres, Passchendaele, Vimy Ridge, were shelled into vast denuded landscapes of craters, mud and filth. The Second World War extended this level of devastation to major towns and cities. Subsequently, the environment itself became a primary, rather than consequential target—the use of the Monsanto-manufactured defoliant "agent orange" in the Vietnam war in the 1960s being the iconic instance. The environmental impact of the destructive force of modern weapons is also more recently evident in the massive fires created by bombing oil fields in the first Gulf war in 1991, or in the destruction in Central Europe across the 1990s up until the environment damage to Serbia that resulted from the "humanist" war to liberate Kosovo. Additionally, millions of hectares of land have been rendered unusable because of the laying of mines in many wars over the last 60 years—which go on killing and maiming people and animals.

Modern war has not only occupied a position of leadership in mass environmental and social destruction, but is also constantly developing more violent means to shatter bodies faster and in greater numbers than ever before. War makes scrap—scrap is the destiny of the war machine; more than this, it scraps bodies and minds. But in contradiction, although steel is the dominant material employed to rip bodies apart, it equally became used to put these shattered body together again: with metal skull plates, bolted and plated bones and the advent of sophisticated artificial limbs.

The destructive impact of modern war did more than just turn bodies and land to waste. The technologies of photography, film and television projected images into the cultures and minds of entire populations, extending the battlefield into the ecology of mind of everyone. Modern war has profoundly changed the psychology of the entire human species. In the sense that war is constantly before us in all its televisual forms (news, film, literature, iconography, fashion, toys, etc.), as well as being a life experience for much of the global population, it has become elemental in the popular imagination. It now exists as a normative feature of the mental geography we occupy. The causal relations between the material and the immaterial dimensions of war ever grow in complexity. Constantly effect becomes cause. For instance, war is enormously destructive of social ecology; it decimates that social fabric of community that binds people together in meaningful and functional relations, and frequently this links to the physical displacement of entire populations. The consequence of the situation of homelessness, fear, insecurity, colonization and psychic destruction is not only that war lives on well after the last battle has been fought, or the last bomb dropped, but equally that the seeds of conflict are ever harbored. In actuality, the impact of war upon ecologies of mind is accumulative. History, by putting conflicts in the past, hides the fact that unresolved conflicts, living on as felt social injustice, travel across time with active agency in the present and determinate force in the future.

As military planners throughout the world are recognizing, there are very dangerous times ahead: global shortages of fresh water in areas of high population, food shortages as a result of the breakdown of land care or as result of crop failure due to climate change, the rising cost of food and conflicts over the ownership and control of scarce material resources are just some of the "flashpoint" issues. It can be noted that two of the most environmentally volatile regions of the world are equally politically extremely unstable—the Nile Basin and the India/Pakistan border along the Indus.

Puffing Billy and the iron horse

War and the war machine were one powerful force driving the advancement, production and use of steel as the first industrial age that began in the sixteenth century was overtaken by the second which arrived in the nineteenth. Another was the iron road—the railway.

Wooden rails, with carts with wooden wheels running on them, were to be found in mines in England and Germany from the sixteenth century. These were later plated in iron (and then called plateways) to enable heavier wagons drawn by horses to be used. Solid iron rails began to be used from the end of the 1760s. In fact, the first iron rails were laid at Darby's works in Coalbrookdale in 1767. Initially rails were cast iron; however, due to its brittleness they were prone to break up. Wrought iron, with its ductile qualities, was used for rails from about the same time and gradually replaced cast iron rails. Due to their brittleness or softness, iron rails had a limited life and were in need of constant replacement. This restrained railway expansion, until the arrival of steel rails later in the nineteenth century. To illustrate: in 1867, 3,000 net tons of Bessemer steel rails were made in the USA; by 1890, the quantity had risen to 4 million tons, with 80 percent of the 200,000-mile rail (320,000 kms) network having steel rails.[19]

In 1805, the first public railway was opened—this was the Surrey Iron Railway, with its horse-drawn carriages. Of course, what really gave the mode of transport its momentum was the steam engine.

The first use of steam power is attributed to the British engineer Richard Trevithick. He progressed from a steam engine (1797) to a road locomotive (1801) and then to a railway locomotive (1804).[20] Others followed—John Blenkinsop patented a railway locomotive in 1811, George Stephenson built his first engine in 1814, William Hedley's engine was completed in the next year. The first proposal to create a railway with steam locomotives pulling wagons to transport people, the Stockton to Darlington line in northern England, was put forward in 1817. The first rail for this line was laid in 1822 and it opened on on September 27, 1825. Travelling at 8 miles per hour (12.8 kilometers per hour) the journey took 55 minutes. Five years later the Liverpool to Manchester line opened. During this period, the most significant locomotive development was Stephenson's "Rocket" (1829). Thereafter, all locomotives were based on advancements from this engine: boiler size (which increased), the overall size of the engine, its engineering, the position of main cylinders, the form of the firebox, the method of steam heating and later super-heating, wheel size, construction and configurations and much more—almost every feature of locomotive technology was designed using the "Rocket" as the normative point of reference.[21]

FIGURE 5.6 George Stephenson's "Rocket" of 1829, the model for future locomotive development. Source: Jacques Payen *History of Energy*, 1966

The locomotive and steam engines in general were constructed to have, and mostly had, long lives. That so many machines, with so many working parts, had lives of between 50 and 100 years is a remarkable testament to the skills of their makers.

Returning to the orthodox history—by 1838 the London to Birmingham line had been opened. This marked the start of a worldwide growth of railways. In Britain in the first 20 years of railways, 6,000 miles (9,600 kilometers) of track was laid and opened. This represented an enormous capital investment for its age, as well as a gigantic civil engineering challenge—the need for cuttings, tunnels and bridges not only made great design demands, but building them involved super-human tasks of manual labor undertaken by an ever-growing number of nations. It was no coincidence that in the early period of railway building Britain's production of coal and iron trebled.[22]

The growth indicated in the early years of railways in Britain was not short-lived. By 1880, 15,000 miles (24,000 kilometers) of track had been laid, and by 1914 this had more than doubled to 32,000 miles (51,200 kilometers)—this increase in mileage was not just linear single track extension, for it also included a good deal of upgrading from single to multiple tracks on major routes. Other European nations, as well as Russia, Africa, India, South America and the USA were also active builders of railway systems from the early years of the technology. Railways were of course one of the prime infrastructural technologies of colonial expansion of European nations and their economies of material appropriation. Equally, the colonial conquest of America was also advanced by it.

Not surprisingly, the performative demands of railways and railway machinery was one of the major factors which drove the growth of the steel industry—as indicated, iron rails were not up to the job; the same limitation applied to wheels, axles and springs required by locomotives, carriages and rail trucks. Thus, between the 1840s and the 1870s many improvements in metallurgy of iron, and the performance of the steel industry, were partly driven by the requirements of locomotive, rolling stock and track construction, and as indicated, the rise of the modern warship.

The "railroad" provided the logistical framework for a national system of communication—it inscribed a system of distribution over a vast land, and as such, it was the key to the economic "opening up" of North America. It made large-scale production feasible, allowing vast qualities of livestock, agricultural commodities and minerals to be transported to their markets. Conversely, the railroad transported manufactured goods to the emergent urban centers of the nation. At the same time, it was a force of colonialism in its own right:

> From 1863 on, armies of imported Chinese coolies laid track, sweating in Nevada's alkali dust or chipping roadbeds into the sides of mile-high Serra cliffs. Starting from Omaha, the Union Pacific Company ... sent its engineers, Irish labourers, teamsters and commissary agents westwards across the plains.[23]

The first railway was established in North America in the 1820s. By 1830, there were still only 23 miles of track (38.8 kilometers). This increased in the next five years to 1,000 miles (1,600 kilometers), in the next ten years to 2,800 (4,480 kilometers), in the next ten to 9,000 (14,400). For reasons connected to the problems in establishing its own iron and steel industry, including the quality of the material produced, during the early period of rail the majority of rails were imported from England, and prior to the Civil War re-rolling rails was big business.[24] The kind of progression indicated continued, so by 1893 America was laddered by five transcontinental lines crossing east to west. The nation's steel industry—and its biggest baron, Andrew Carnegie—was built on the back of this enterprise. By 1900, the US railway network consisted of nearly 260,000 miles (416,000 kilometers) of track.[25] To give some sense of the amount of iron and steel involved, consider that in 1874 a tonnage was calculated at 150 tons per mile (this was based on rail and rolling stock weight). While the strength of steel lowered this, if we bring a lowered weight (say 120 tons per mile) to the 1900 track distance, we arrive at the extraordinary figure for the time of 31,000,000 tons of steel in the system.

For all the engineering and environmental impacts of the creation of a railway network, for all its economic significance, and for all the competitive aggression between the world's major builders of locomotives (Britain, France, the USA, Germany and Belgium), this was not its greatest consequence. By far the most dramatic impact of the railway was cultural—it brought new futures into the world, and took other futures away (it *defutured*).

The railway changed time. Its speed overcame distance, it connected what had previously only been connected for a very few travellers and thus transformed cultural exchange forever. This change was felt at the time, if not fully understood. As Eric Hobsbawm noted, railway mania swept the country and brought almost the entire population into proximity with industrial society.[26] It also laid down a completely new infrastructure for the distribution of goods.

As navies modernized, the sheer mass of metal in the military machine decreased. This reduction was offset by the rise of mechanized armies and armored fighting vehicles. Either way, the war machine is always scrap waiting to happen. A similar relation exists in the field of transport. The railway system has significantly shrunk worldwide; at the same time, and with much greater negative environmental consequences, the use of motor transport constantly increased, and still increases. The flow of "scrap waiting to happen" is ever faster. This "development" also corresponded to a proliferation of freeway building which started in the 1930s and continues.

Freeway construction, with its vast appetite for steel reinforcing mesh and bar, links us to what has been a continuously growing market for steel from its inception—the use of steel in built structures. Freeways, urban infrastructure, commercial, institutional and domestic steel-framed structures are all linked to a constant and massive global trend: urbanization.

Infrastructure

Iron was used to construct infrastructure from its earliest days. It was used, for instance, in bridge building (as chains and linked rods) in China from the sixth century and then later in building suspension bridges from the sixteenth century onward. The first European suspension bridge was built in Britain over the River Tees in 1741. The very first iconic form of the modern "industrial revolution" was of course the iron bridge made at Abraham Darby's works at Coalbrookdale and erected in the village (which was to become known as "Ironbridge"). This structure took three years (1776 to 1779) to cast and build.

Bridges became some of the earliest, most impressive and largest feats of the use of iron and steel. Two British examples illustrate the point: the suspension bridges built across the Menai Straits by Thomas Telford, with seven short spans of 16 meters and one long one of 176.5 giving a total of 288.5 meters—it was completed in 1825; and Brunel's Royal Albert Bridge, a suspension bridge of two spans of 930 feet (almost 285 meters) built across the Tamar at Saltash, and opened in 1859. A much earlier but less visible infrastructural application was the use of cast iron pipes, which were first used to bring water to the Palace of Versailles in 1664 (some of which were still in use in the 1960s).[27]

Besides bridges, many other now familiar steel structures were to become part of the cityscape and landscape. Different kinds of towers, pylons, silos, cranes, weigh-bridges and fences all marked the visible presence of steel. Besides pipes, there was also an expansion of invisible

FIGURE 5.7 The first iron bridge in the Western world—built at Coalbrookdale, Shropshire, England between 1776 and 1779. Courtesy of Kippa Stock Photography

applications of steel, as reinforcement in concrete for roads, runways, tunnels and numerous other built structures.

Iron and steel frames and structures

Buddhist "architects" used iron as framework in the building of pagodas—the oldest surviving into the modern age being the 13-storey Yü Chhüan Ssu temple built at Tang-yang in Hopei in the eleventh century.[28] The first iron-framed multi-storey building in the West was constructed in 1792–3 at Shrewsbury Mill, England.[29] While there were a number of iron-framed structures built in the nineteenth century, the one which caught the world's eye and imagination was Joseph Paxton's Crystal Palace, created for the Great Exhibition in London's Hyde Park in 1851.[30] Although volumes have been written on the event, the structure, its builders (Fox and Henderson) and Paxton himself, the real level of attainment is still hard to grasp. The perception of the attainment is in large part obscured by the volume of massive structures constructed since then. But if one makes the effort to try to consider its scale, form and technology in its context, it takes one's breath away.

The iron architecture of the Crystal Palace came at a moment when iron was just about to give way to steel; nevertheless it was, and in some ways still is, a major statement in the script of the future.

Its architectural form rested on the skills of the structural engineer; its aesthetic was predicated upon the transparency of its mode of design and assembly (both of these features make it very contemporary); it employed industrial rather than craft methods of manufacture (it was architecture by iron-making, forging, casting, machining); it used prefabricated methods of assembly via the production of standardized interchangeable parts; and it was designed for disassembly/reassembly (and to a scale that is still very relevant to contemporary building and construction practice).[31] While the Crystal Palace structure was over half a kilometer long, it was designed to be endlessly extendible.

The Crystal Palace took design and construction with iron and steel into a new era. As its prominence in the history of design, architecture and art indicates, it became a type form.[32] This was evident through the speed of its international influence. Many other cities wanted such a structure of their own—hence the structures of the New York Exhibition

FIGURE 5.8   Galerie des Machines constructed for the Paris Exhibition of 1889. Source: Leonardo Benevolo *History of Modern Architecture* Volume 1 London: RKP, 1971; trans. from *Storia dell'architettura moderna*, Italy: Giuseppe Laterza & Figli, 1960

of 1853, the Munich Exhibition of 1854 and the Paris Exhibition of 1855. The Crystal Palace also generated a whole manufacturing and export industry of smaller prefabricated structures, which were mostly exported to "the colonies." At the outbreak of the Crimean War (1853–5), an entire prefabricated barracks was exported.[33]

In sum, the Crystal Palace was a proto-paradigmatic structure of: mass production building methods of assembly and disassembly; building systems organization (logistics, labor and project management); building as spectacle; and of multi-life building (materially in its use at Hyde Park and later, re-assembled at Sydenham, immaterially as a typeform of iron, and a few decades later, steel-framed buildings).

While iron structures originated in the birthplace of the Industrial Revolution, they came into their own in the New World. The first event that triggered this was the fire of 1871 that almost completely gutted Chicago (which by then had a population of over 300,000 people). The second was the construction of the iron/steel-framed skyscraper in New York 1883—which prefigured all tall buildings with load-bearing walls in that city and the rest of the world.[34] These two events converged.

The first all-steel-framed building was built in Chicago—the Rand McNally Building (1889) designed by one of the most famous of the city's architectural practices, Burnham and Root. The impetus to use steel came from the fear of fire and the desire for tall buildings. While many years of applied research were needed in order to fully realize the potential of steel to reduce risk *of* fire (and risk *in* fire), it should be recognized that in this context steel was used for its symbolic as much as its practical value.

The Rand McNally Building employed steel beams and riveted steel members in column fabrication, a type of construction first used for bridges. These construction methods established a base for all subsequent skyscrapers. As Hitchcock says: "Steel construction of the American Type, with the internal skeleton carrying a protective cladding of masonry, has gradually spread since the opening of the century to all parts of the world that produce it and can afford to buy structural steel."[35]

Steel-framed high-rise buildings have not only transformed the nature of urban development and land economics, especially in the commercial areas of cities, but also the character of building services and the environmental impacts produced by building construction and use. On the last point we should note the introduction in the Haughwout Store in New York of the first passenger elevator in 1857 (as developed by Elisha G. Otis) and the first effective air-conditioning plant (developed by W. H. Carrier) in a printing plant in New York in 1902.

This was the beginning of an era of dramatic increase in *engineered* environmental impacts. Consider the combined greenhouse gas emissions from the embodied energy of steel (the accumulative sum of invested energy from material extraction to end-product manufacture), from the energy uptake of building mechanical services (especially lifts, air-conditioning and artificial lighting) and later, from electronic office technologies. Added to this is the growth of steel building products (from roof sheeting and guttering to prefabricated computer floors with in-built cable trays and channels). Thus, steel in the modern world both exists within, and has been generative of, material ecologies that have had enormous environmental impacts. How these came to be recognized and how they are being sought to be reduced will be the subject of following chapters.

At the same time these developments also expanded the commercial opportunities for architects, engineers, specialist contractors and building contractors—evidenced especially by the growth of large American cities. Now, in the early decades of the twenty-first century, the "tide has turned." The USA has been de-industrializing: Detroit had a population of 2 million in the 1950s, now it is 600,000. In the same period Cleveland shrunk from 1 million to under half that. A similar story of steel-making or steel-using cities across the US can be told.

As so much of the content of this and earlier chapters makes clear, technology ended the hegemony of the natural, but, in turn, technology became naturalized.

The environmental changes created by war and industrialization; the invention of materials (especially iron and steel); the great machines of land and sea; the vast structures built by the hand and tools of man; labor in light, heavy and service industry; lives lived in the shadow of fear from weapons of mass destruction; the technologies of mass communication in war or peace; lifeworlds populated by ever more valorized and manufactured products; the impacts of industrial work and war on the body and mind—the listing of the transformations of modernization has the ability to ever expand, but even from its brief beginning we can see, if we pause to consider, that there is now no possibility of establishing any clear division between the natural and the artificial. The ecologies of steel are thus not supplementary to a primary ecology, but elemental.

Notes

1   We have already noted that weapons and armor were not only among the primary products of early Chinese iron-makers, but they also

provided the impetus to advance the organization of industrialized labor. Weapons could also carry a great deal of cultural status—to own a sword of Damascus steel in the first century was to be numbered among the extremely privileged.

2   See Arthur Ferrill *The Origins of War* London: Thames and Hudson, 1986, pp. 67–71.

3   Ibid., p. 86.

4   Ibid., p. 149.

5   Ibid., p. 215.

6   The Roman Empire spanned a period of two decades prior to the first millennium and five centuries after.

7   Harry Scrivenor *History of the Iron Trade* London: Longman, Brown, Green and Longmans, 1854, p. 30.

8   Robert O'Connell *Of Arms and Men: A History of War, Weapons and Aggression* Oxford: Oxford University Press, 1989, p. 112.

9   This weapon was used with lethal consequence in the Danish war of 1864, the Austro-Prussian war of 1866 and the Franco-Prussian war of 1870. See O'Connell *Of Arms and Men*, pp. 204–11.

10   Merritt Roe Smith (ed.) *Military Enterprise and Technical Change* Cambridge, MA: MIT Press, 1985 pp. 8–9.

11   David Lyon *The Ship: Steam, Steel and Torpedoes* London: HMSO, 1980, p. 5.

12   Ibid., xx.

13   While the concept of the submarine is attributed to Leonardo da Vinci and the first example to David Bushnell (a Yale undergraduate) in 1773, the first powered vessel is credited to John Holland, an Irish American, in 1881, which took 15 years to fully develop. The first true ocean-going naval and deadly submarine was the German U-19 of the First World War.

14   The main battleship armament in 1873 was the 35-ton, 12-inch gun (with a shell weighing a third of a ton); by 1878 this had been displaced by the 80-ton, 16-inch gun (with a shell weighing three-quarters of a ton).

15   The Dreadnought's speed of 21 knots was not only 2 knots faster than any other battleship, but, unlike its competitors, it could maintain this speed for many hours (this was possible because it was powered by a new technology, the steam turbine). It was armed with ten 12-inch guns, twenty-seven 12-pounders and five torpedo tubes. Defensively, the ship carried 5,000 tons of armor, which was generally between 2.5 and 4 inches thick. In specific target areas, like the conning tower and the face of the gun turrets, it was 11 inches thick. Some idea of its size is conveyed by its crew of more than 700 men. The ship's speed of construction—just one year and a day—was another reason why it was so feared.

16   The battleships became the norm against which all other fighting ships were ranked: battleships, heavy armor, heavy long range turret guns (first introduced in 1860), slow; cruisers, light armor, heavy turret guns, light guns, fast; submarines, unarmored, armed with torpedoes, single light guns, medium speed on surface, slow below surface; destroyers, very lightly armored, medium turret and light guns, fast; frigates, no armor, light guns, fast; motor torpedo boats (MTBs), unarmored, torpedoes, very light guns, very fast.

17   The development of armor plate was prompted by several factors all coming within a few decades of each other: the introduction of explosive shells, the increase in the size of guns, shell velocity of shells and shell accuracy (due to rifling), plus the introduction of the torpedo. What especially drove its development was the arrival of the steel armor-piercing shell in the 1880s. Armor plate was created by a variety of alloys and processes. The most common alloys were vanadium (a lightweight metal discovered in 1830 that, when combined with chromium and added to steel at a level of 1 percent, created a tough high impact-resistant alloy—which had many applications beyond armor plate) and manganese (a metallic element that is softer than iron and non-magnetic when alloyed with steel, which was first done by Sheffield steel-maker Robert Hadfield in 1882—manufactured from 1892, it was widely used in the First World War for spur armor plate, shells, tank treads and helmets. Additionally, American engineer, Hayward Augustus Harvey devised a face-hardening process whereby carbon was applied to the face of a nickel alloy steel plate at very high temperatures, which was then tempered for an extended period. Krupp's hot-gas-tempered armor plate was based on Harvey's methods. Yet another steel alloy in the early phase of the armor plate manufacture was a chromium/ nickel combination. The chemistry of contemporary armor plate draws on an increasingly complex range of alloys.

18   The motor torpedo boat was first introduced in 1876 by the British shipbuilder Thornycroft who built *HMS Lightning*; these boats were powered by locomotive engines. Torpedoes were created in the same period, and were, for instance, successfully used by the Russians against the Turks in 1877–8.

19   D. A. Fisher *The Epic of Steel* New York: Harper and Row, 1963, p. 125.

20   The steam engine used in locomotives, steamships or industry seems to be a technology of the past, yet it may well be a technology of the future. Perhaps such a view is found in the very nature of the technology itself. In its most overt form this is seen in statements like this—"The steam locomotive is unlike any other machine in one remarkable respect. Its various elements all interact, rather as they do in the body of an animal." See O. S. Nock (gen. ed.) *Encyclopaedia of Railways* London: Octopus, p. 149.

21   This information (and much else on the topic of railways) is drawn from Nock *Encyclopaedia of Railways*, p. 154.

22   Eric Hobsbawm *Industry and Empire* Harmondsworth: Penguin Books, 1968, p. 71.

23   Bernard A. Weisberger (ed.) *The Age of Steel and Steam* New York: Time Life Books, 1964, p. 30.

24   Between 1840 and 1860, 60 percent of rails were imported from England.

25   Nock *Encyclopaedia of Railways*.

26   Hobsbawm *Industry and Empire*, p. 110.

27   Leslie Aitchison *A History of Metals* Volume 2, London: MacDonald and Evans, 1960, p. 438.

28   Joseph Needham *The Development of Iron and Steel Technology in China* London: Newcomen Society, 1958, p. 20.

29   Alan Ogg *Architecture of Steel: The Australian Context* Canberra: Royal Australian Institute of Architects, 1987, p. 58.

30   The structure was subsequently moved to Sydenham in 1852–4, where it stood until, ironically, it was destroyed by fire in 1936.

31   Consider the structure had: 330 iron columns, 2,224 girders, 1,128 bearers for the gallery level, 54.5kms of guttering and 328kms of sash-bars, plus a mountain of glass.

32   See, for example, the influential Leonardo Benevolo *History of Modern Architecture* Volumes 1 and 2 (1960), English translation, London: RKP, 1971, and the multi-edition Henry Russell-Hitchcock *The Pelican History of Art* Harmondsworth: Penguin Books, first published 1958.

33   Hitchcock *The Pelican History of Art*, p. 188.

34   The nine-storey Tribune Building of 1873 and the ten-storey Western Union Building of the same year, both in New York. The first skyscraper was in fact the ten-storey (with two added later) Home Insurance Building designed by W. Le Baron Jenney—interestingly, while starting out building a frame that took all load bearing off the walls (that nobody had done before) with cast iron columns and wrought iron floor beams, by the time the building reached the sixth floor Jenney received a letter from their supplier of iron, the Carnegie-Phipps Company of Chicago, which was to alter the course of history. The letter informed Jenney that the company was now rolling steel beams produced by the Bessemer process, which were proposed to be substituted for the iron ones. While the columns continued to be iron, the use of steel beams marked the first use of structural steel in the first skyscraper. See Col. W. A. Starrett "The First Skyscraper" in Lewis Mumford (ed.) *Roots of Contemporary American Architecture* New York: Dover, 1972, p. 237. The word skyscraper came from the name of the highest sail on the clipper ships.

35   Hitchcock *The Pelican History of Art*, p. 351.

FIGURE 6.1 Machine Shop, 1950s, UK. Photographer unknown. Authors'
collection

6 TECHNOLOGY AFTER THE MODERN

A dopting multiple perspectives on how steel is viewed as a material, the manufacturing trends of the industry and the shifting symbolic significance of steel will be our concern here. In common with the last few chapters, we will start with general observations followed by a more focused look at specific examples to support the overall argument.

Steel at the end of "The age of the world picture"

By now, it should be very clear that steel occupies an unrivalled place among materials. No matter who or where we are, it is almost impossible to imagine how our world would function without it. Besides those obvious and historically identifiable cultural, environmental and economically transformative technologies and products that depended on steel for their manufacture—like railways, battleships, skyscrapers, motor cars—there are an almost incalculable number of steel components embedded in the fabric and functional operation of the entire made world.

At the start of the twenty-first century steel ranks only second after concrete as the most prolific material manufactured by human beings. The comparison ends there, for as indicated the diversity of the applications of steel from small precision components through to gigantic built structures, and the number of industries dependent upon it in whole or part, put it into a different frame of reference than concrete. Yet the lifespan and the environmental impacts of these materials are increasingly being compared.[1] Viewing these materials independently obscures the significance of interdependent relations—reinforced concrete is an obvious example of a material whose performance depends upon a designed relation between two materials, steel and concrete.

The ecology of a manufactured material, as we have been at pains to show, is always plural. For all its visibility, steel predominantly exists in a condition of absolute taken-for-grantedness—its actual omnipresence passes us by. While we can point to objects made of steel, what is less obvious is its embedded function in the technological substrate upon which a good deal of late-modern life depends. The biophysical environmental cost of steel's production, the ecologies it has destroyed, altered or constituted, the cultures of its industry, and the relation between it and our future are by no means instantly visible. Mostly such concerns about steel are restricted to the industry and environmental watchdogs, and focus on its manufacture. Unless there is an overt problem for a specific community, the environmental impacts of steel seldom arrive as the concern of public consciousness. Notwithstanding this, there are tensions between what steel enables, what it costs in the broadest sense and the development of informed judgment about its use. In the global circumstances of still increasing unsustainability, the question of "how to" think about steel—and the material world in general—is of growing importance for us all. Certainly, such questions cannot just be left to technocrats.

In recent times, the primacy of material production has been challenged by the rise of sign economies, wherein image and brand become forces of production in themselves rather than simply the means of bringing identity to already made products. While first evident in America in the 1930s with the deployment of streamlining as product styling to generate consumer demand and economic recovery, the "sign economy" as a global force did not mature until the 1950s. This occurred with the arrival of popular culture products in the commodity domain of "youth culture" and with the growing brand power of corporations created by the convergence of television, mass audiences and advertising. The second challenge to the material economy arrived in the 1990s with the coming of large-scale share trading in internet economy stocks and e-commerce. The basic proposition of this economic activity was that it was possible to generate wealth immaterially—wealth did not have to depend on actual transformations of matter in "the real" material world. Leaving aside the problems and implications of this thinking for the moment, what we do need to recognize is that one of the effects of this rising discourse of immaterialization has been a displacement of the significance of the material in the minds of many people. Effectively, made materiality, its environmental impacts and the environment itself have been downgraded. Moreover, many dirty manufacturing activities get pushed out of sight—not least to "newly industrializing nations." In contrast to the status of some material activities being diminished in the already industrialized world, the demand for materials in the newly industrializing nations is still rapidly growing—not least through a convergence of ever more urbanization, growing industrial output and a constantly expanding desire for modern and stylish manufactured commodities.

The concealment of materials by aesthetics and the "immaterialist" rhetoric that downgrades the status of materials are contradictory characteristics of the postmodern age, and elemental to its unsustainability. Increasingly, more human beings desire a secure, sustainable environment and future, but they also desire lifestyles, resources and goods that fundamentally negate what has to be sustained. To counter this situation two things need to happen: first the contradictions have to be made explicit; and second, another discourse of the material begs to be created that places "materiality" within regimes of resource conservation underpinned by cultural as well as economic and environmental values. Materials practices have to go beyond manufacture and recycling into new activities like re-materialization and other redirective practices that we will be outlining. These practices are predicated upon a far more highly developed understanding of, and intervention in, the relations between material production, the making of things, time and the formation (and de-formation) of worlds. Above all, values have to be redirected from the quantitative to the qualitative.

Seeing steel

Binary divisions between the natural and the artificial, biological and mineral matter, plants and chemicals all still figure in the "common sense" Western way of seeing and describing the "order of things." However, this is at odds with how contemporary biological sciences, physics, chemistry and critical theory understand the complexity of matter and its transformation over time.

Positivistic common sense directs us to "see" matter as static. We often have little sense of how a material (made or given) was formed or of how variable conditions produce molecular change. Moreover, while we posit matter as existing at an atomic level, we negate its temporal condition of differential flux, which is to say that all matter changes over time, be it at enormously varied speeds—this due to both internal or external factors. So said, the very concreteness of that which we take to be certain is lodged in an unseen flow of time and movement that so often defies how a material has been classified. Steel is one of the more legible of materials. It arrives from an industrial process, which can be seen as a cycle: ore extracted from earth, the making of iron then steel (with perhaps a cycle of remaking) to its eventual degeneration into iron oxide and its return to earth—this is an inevitable process. However, this cycle is not "natural"; its un-naturalness was prefigured and prompted by the appropriation of a "natural phenomenon"—meteoric iron (which existed millions of years prior to human life). Thus, prehistorically, the story of steel, a material deemed to be deeply inscribed in the artificial, began with the supra-terrestrial "natural" chance production of iron, its discovery by human beings, its slow passage to utilization and eventually, its simulated manufacture.

De facto, we have just acknowledged that materials and technologies always exist in an entropic economy—they are discovered or invented, they develop, they peak and then are destined to eventually lose their utility and their economic and symbolic value. The relation between timber, coal, iron and steel illustrates various moments of this process. Theory and history have to stand in for our inability to view slow-moving material changes as they occur over many decades, centuries or millennia.

Seen in time, steel looks like a primary material of modernity that, while having a long afterlife, is now past its prime. While steel will undoubtedly go on being produced and used in vast quantities into the future, it is no longer the hegemonic material of modernist manufacture. This is not because of diminished usefulness but because there are now other materials and technologies with which it has to compete, both performatively and symbolically. The steel industry is partly aware of this, and, at its leading edge, is contesting the ground by exploring ways in which the industry and the material could reinvent itself. A good deal of contemporary innovation in steel metallurgy is driven by this imperative in the face of competition from carbon fiber; very hard, extremely durable ceramics; and complex composites. Besides economic factors, what will also increasingly play a decisive role in the fate of established or new materials will be the extent of their environmental impacts in production, use, recovery or disposal. To claim that steel is no longer a hegemonic material does not suggest that it does not have an enduring future. Certainly demand drivers, like the projected rates of urbanization, will ensure that the construction industry will go on using vast amounts of steel for a long time. Yet the symbolic power of steel will increasingly come under threat from other materials. It will be harder for steel to hold center stage in the theatre of the image. In this respect, its future is dependent as much upon the immaterial factors of remade symbolic forms and market demand as on material production factors.

The time of the geography of modernity

Every moment in the history of a technology is a registration of a moment in time. Technology's progression is usually treated teleologically and laid out across an evolutionary line. Yet technological transformations (and many other changes) are seldom concurrent. As our account has attempted to illustrate, the historical development of a technology is uneven, with the crude and complex mutually coexisting at both the same and varied moments across culture and geography. The narrative ordering of history writing usually negates the contradictory and untidy character of events as they happened. We have tried to counter this, showing the development of iron and steel to have occurred in plural time. At its simplest, if we view steel-making globally at "this moment of time" what do we see? Clearly in some parts of the world steel is still being made inefficiently (in terms of labor, material and energy) in batches by open-hearth blast furnaces much as it was in 1900. At the other extreme, the most advanced electric arc furnace steel-making is edging towards a fully automated continuous process whereby scrap or directly reduced iron flows in and marketable steel rod, strip or sheet product flows out.

The plural nature of time is more complex than this brief example can indicate. It maps onto the difference of socioeconomic circumstances, working conditions, health and safety, lifestyle, dreams and desires that are linked to varied circumstances and geographies. Equally, as we have shown in previous chapters, steel was the material of the technologies that enabled the further expansion of imperial power of the first industrial nations. As such it was a material of modernity that was directly employed to intervene in time. It now exists as an active agent in extending the modern and the postmodern.

A snapshot of the globalization of steel-making, the trade in steel products and the symbolic alterations in the meaning of steel in the last 50 years will be helpful here.

Steel times

The dynamic cyclical relation between demand, industrial production and destruction meant that the Second World War was good for the steel industry. Certainly the wealth, volume of productive output and freedom of the US industrial infrastructure from bombing were major factors in the "free world" winning the war. Thus the war provided a major boost to the US steel industry, pushing it towards global dominance for several decades. In fact the steel industry was at the core of the US economy's post-war expansion. It fed the rapid growth in the automotive and domestic appliances industries, the impetus of post-war reconstruction plus cold war military modernization.

However by the 1960s and more intensively during the 1970s, the US steel industry was being overtaken by new global players.[2] Japan was building a large modern steel industry, with almost three times more investment drawn from net sales than the US was putting into research and development.[3] Ironically, in an age when the ideologies of "modernization" and "development" were aggressively promoted by the US around the world using the United Nations as the instrument of delivery, the US steel industry failed to modernize itself. Meanwhile, "just-in-time" methods of inventory and "integrated producer" methods (like computer-aided manufacturing) arrived, initially from Japan and then Europe, which led to industrial modernization.[4]

It would be erroneous to view the changes in the period under consideration as only technological, economic and political. As we saw in Chapter 4, when reviewing steel and modernity in China, steel was symbolically mobilized with the "Great Leap Forward" campaign of communist China during 1958–9. As said, "this campaign elevated iron and steel as iconic materials for China, and they have remained in this position ever since." As with the USSR in the 1930s, China employed steel as a sign within industrial production, which itself was one of the key agents in the creation of the communist state. Equally, steel was deployed by US capitalism as a symbolic material of progress (seen, for example at the New York World's Fair of 1939, in which steel was presented in the form of iconic structures like the US Steel building.[5] Notwithstanding the arrival of 1960s environmentalism, or the recognition since the late 1980s of pending environmental crisis from global warming, neither the rhetoric of "development" nor "globalization" evidences any real recognition of the imperative of sustainability as the essential basis for a viable world order. Social justice, economic and environmental responsibilities are still not understood as practical and ethical necessities.

In the 1960s "developing countries" produced only 6 percent of the world's "crude steel"—this being a significant advance on earlier volumes.[6] By the 1980s the quantity had doubled and a new pattern was emerging. More and more "developing world" steel-makers were entering the world market. The percentage is now of the order of 30 percent, although the situation is more complex as the lines between "developed," "developing" and "newly industrialized" get more blurred and more a matter of internal, rather than international division. To illustrate the point, Brazil, the Republic of Korea, India, Mexico, Turkey, are all major steel producers, as is China which is the world's largest.

One sobering fact and one trend evident in the recent pattern of growth stands out: in the period 1960–2000 world production of steel almost quadrupled, from its 1960 base of 200 million tonnes—it now exceeds 1,500 megatonnes.[7] The trend for the quantity of steel produced by "developed" world steel-makers is either a slight increase or fall, but with China now being the largest overall producer. These trends do not rest well with the growing imperative for the industry to play a role in establishing conditions of increased sustainability. This role would involve structural changes in order to deliver a significant overall reduction in negative environmental impacts. For this to happen production has to fall, environmental control has to increase, roles have to change and wealth, including commonwealth, has to be created. None of this can happen without changing what "steel" signifies, how it appears and how it is used.

While some site-specific impacts of steel-making have dramatically improved in comparison with the past (when production methods were cruder and more environmentally harmful), the overall situation is still negative because of the multiplier effect of the massive increase in the volume of steel made today. Defuturing is a structural feature of a steel industry. Although partly embracing the rhetoric of sustainability (and "cleaner production") it is still deeply implicated in "sustaining the unsustainable." Even though this situation will continue, the imperative remains to create an ecology of steel with lower impact.

While no magic solution, the electric arc furnace (EAF) has made a significant contribution the environmental performance of the industry. From their introduction in the early 1960s to today, EAFs have become more efficient. On a comparative basis per ton of steel, CO_2 emissions from a scrap-charged EAF are claimed as being just 25 percent of those from an integrated steel works[8]—with the energy employed normally coming from burning fossil fuel. At the same time, the number of integrated steel works has decreased and the most inefficient modern means of making steel, the open-hearth blast furnace, is also in rapid decline. However, the size and output of those remaining integrated steel works (and the few still being built) has increased (in the year 2000 there were fewer than 10 percent of the number of integrated steel works in the USA than in 1900, but these were producing 20 times more steel daily than any of the works a century ago).[9] Meanwhile the number of EAF serviced mini-mills has constantly grown.

While opening a Pandora's Box of complexity, we really do not have a usable sense of impacts until we build a cumulative picture. Doing this cannot presume a final truth will ever be reached, rather the objective is to gain a picture that can assist understanding and judgment to usefully inform action. Lack of certainty is in fact not alien to science, but a condition of normality as it commences a task first of all by a leap of faith (at one pole we can say that solipsistically, reason can only prove what reason claims to be true by reason; at another one can take a specific example, like "life-cycle analysis" which commences from unscientific assumptions and thereafter proceeds to utilize scientific data to author a trope).

Steel and the dominant trends of now

A momentum towards improvements and efficiency in manufacturing methods and in the quality of metal produced is now intrinsic to the steel industry, not least because of the competitive nature of the industry.[10] This has meant that basic operations have moved to locations where raw material and/or labor costs are cheaper as well as utilizing advanced technology that completely eliminates live labor. Again this is a trend in keeping with a more general pattern of capital striving to maintain and increase profitability.[11]

These processes need to be understood as incremental steps towards the end of all batch production methods and the establishment of fully automated synchronous steel-making.[12] EAF has been taken up from the extreme geo-economic poles of the steel industry—it has a high use in the United States, as the world's wealthiest economy, and in the "newly industrializing" nations where steel-making is employed as a path to enter (and a sign of arrival of) the industrial product sector of world trade.

Mini-mills and EAF technology were attractive to "newly industrializing" nations because of affordability. For most "newly industrializing" nations the cost of an integrated steel-making capability was prohibitive. Thus the widespread introduction of EAF steel-making has led to a very competitive environment—for example, steel-making in Mexico doubled over the 1990s. Mexico has one of the most sophisticated mini-mills technologies in the world. The steel industries of "newly industrializing" countries thrive because of the market demand for steel from their own rapid urbanization. Nowhere demonstrates this more overtly than China. Its size plus the speed and extent of its urbanization has accounted for a good part of the annual growth in steel demand. The volume of steel produced is used in construction—for example, in 1997 it was around 43 percent of total output (which was 108,588.7 million metric tons).[13] While China's steel output has constantly grown, the volume produced by the large state-owned integrated steel plants declined and the number of small and medium-sized quasi-private producers using EAFs and mini-blast furnaces expanded.[14] There is another factor in the spread of EAF technology. The larger "newly industrialized" steel-making nations like Mexico, Brazil and Korea are playing a significant role in extending the industry in less developed economies—usually by forming joint ventures, with mini-mills as the primary building block.

From an emissions-reduction point of view, scrap is far superior to direct-reduced iron (DRI) as furnace feedstock. This is even more the case in relation to hot metal from integrated steel works blast furnaces. Hot metal from charcoal-based mini-blast furnaces is, however, a more viable and still underdeveloped option, provided the charcoal is sustainably manufactured. From the perspective of steel quality, iron substitutes and mini-blast furnace iron are a good means to counter the "tramp" content in scrap that lowers the quality of steel.

Learning and defuturing

The "future of steel" has to be engaged within a planetary rather than in an industry context.

To a very significant extent, what we have been setting out to promote is a sensibility that is better equipped to think such change as sustainment and act toward its advancement in very real ways. Learning what is unsustainable—that is, learning what takes futures away (what "defutures")—is a prerequisite for advancing sustainment. Without this knowledge, there is no possibility of discovering what already sustains and what means of sustainment need to be created. This learning requires looking back to look forward, looking for future inscriptive forces in the afterlife of the past carrying forward into the future. In the end, the most successful steel companies will be those that embrace a radical transformation of what they are and do as minders and managers of resources. Clearly the actual implications of making the steel industry "sustainable" are profound. They require making a very clear distinction between "creating the ability to sustain" and "sustaining the unsustainable." For the steel industry to become a means towards sustainment, and thus for its wealth, viability and advancement to flow from serving this end, totally new ways of thinking and acting are essential. The seemingly impossible has to be attempted and achieved.

The scale of the challenge and the trans-generational implications of sustainment need to be grasped. The agenda of sustainment requires a broadening of focus well beyond the remit of one's existing knowledge, practice or industry. Equally, it requires a facility to move from understanding "the big picture"—the significance of anthropocentrism—and the detail of a specific context. In all our differences, we are only still at the very earliest moments of this massive, pressing and mostly unrecognized re-directional project, the fate of which still hangs in the balance.

It took human beings many thousands of years to create civilization, explore and map the planet and establish the foundations of human knowledge; it took several hundred years to make a partly modern world on this planet; and it has taken several decades for a minority to realize the error of our auto-destructive mode of occupation (manifest not just in the ways we make, build and dwell, but also in the way all these activities are prefigured by the way we think and design).

The challenge of sustainment is set against this backdrop. It centers not on our exercising a stewardship of "the world" but rather a stewardship of our embodied (life roles) and disembodied (institutionalized) selves that recognizes what needs to be made, conserved or unmade. The essential project of sustainment fundamentally confronts the anthropocentric essence of "us"—the locus of the unsustainable. This anthropocentric quality of human being, while not able to be transcended, can be instructed (taught and/or ordered) on the basis that "self-interest" has now become indivisible from responding to "the interest of human and non-human others." What we have now then is an "extended responsibility of the self" which can act to open a future rather than defuture.

Having created conditions in which the future is no longer an assured "event," our responsibility now extends to the making of time, which means acting to identify and destroy the things that take the future away (defuture).[15] This process of identification is one that begs to be learnt—in large part, it is an intellectual skill, which forms part of the "discipline" of "extended responsibility." The entire project of this text has been informed by such thinking. The archaeology of the past that has been undertaken was not done as an historical, academic exercise, but as a practical one connected to "the design of the future." It is in this setting of the significance of the (anthropocentric) self to the "world" and of the past to the future, that *futures* of the steel industry are considered.

Notes

1   One of the driving forces of "life cycle assessment" (LCA) has come from materials producers who seek to compare the environmental impacts of one material with another. This is done, for example, by using a common comparative reference such as the composite sum of energy embodied in a certain amount of material (expressed as the embodied energy of a specific weight) over its entire life cycle. For example, the Australian steel industry developed life cycle assessment tools in response to the timber industry's claims of timber-framed project homes having less embodied energy than steel-framed houses of the same type. Although methods employed by LCA are scientific, the boundary-settings and opening assumptions of the exercise are very often subjective.

2   The world economic order of the 1950s and 1960s was shaped by two political currents: the Cold War and the agenda of world modernization— this at a time when the European nations were dismantling their old empires. It was during this period when the idea of three worlds was

created: "first world" (modern Western capitalist), "second world" (Soviet bloc) and "third world" (colonized or newly post-colonial poor nations).

3   The actual figure was 0.6 percent in the US (of which most went into product development) in contrast to 1.6 percent in Japan (of which most went into new technology research). Dennis A. Ahlburg et al. "Technological Change, Market Decline and Industrial relations in the US Steel Industry" in Daniel B. Cornfield (ed.) *Workers, Managers and Technological Change* New York: Plenum Press, 1987, p. 237.

4   Michael J. Piorre and Charles F. Sabel *The Second Industrial Divide* New York: Basic Books 1984, p. 209.

5   This event, as a symbolic marker of the overcoming of "the depression," can be claimed as the most significant of the many World Fairs of the twentieth century. See Joseph Cusker et al. *Dawn of a New Day: New York World's Fair 1939/40* New York: Queens Museum/New York University Press, 1980.

6   Ibid., p. 208.

7   See World Steel Association: http://www.worldsteel.org/media-centre/press-releases/2012/2011-world-crude-steel-production.html

8   The electric arc process was actually invented in 1878 by Sir William Siemens, but not put to practical use until 1886, by Paul Heroult in France (to make aluminium). While usefully employed in making alloys and specialist steel during the First World War, the size of the furnaces were small with limited output. The technology did not gather momentum until the rise of the "mini-mill" in the early 1960s. See W. K. V. Gale "Origins and Development of Small-scale Steel-making" in R. D. Walker (ed.) *Small Scale Steel-making* Barking (UK): Applied Science Publishers, 1983, pp. 1–19.

9   Bryan Berry "A Retrospective of Twentieth-Century Steel" *Iron Age-New Steel* November 1999 (web edition, p. 4/18).

10 The introduction of BOFs spelt the end of open-hearth furnaces (although the death has been slow and lingering). BOF technology had four advantages—speed (the melting time of a heat of BOF at 45 minutes is ten times faster than open-hearth); significant fuel saving; dramatic savings in labor (*New Steel* cites William Hogan's claim that BOF in 1999 requires 1,000 fewer manpower hours per ton than an open-hearth furnace required in 1920); and the capability of producing many grades of steel. The one advance of the open-hearth was flexibility of the volume of scrap able to be used in the charge, which could be between 20 and 80 percent. Although BOF is not as fast as the Bessemer process it displaced, it delivered a much greater capacity (most BOFs are of the 100–250 metric ton range, which is 6.5 to 16.5 times greater than a Bessemer converter) and offered much better control over quality (the violent chemistry of the high speed steel-making of Bessemer—around 12 minutes—restricted the ability to conduct metallurgical tests and to carry out adjustments if

necessary). Continuous casting was a process created in Germany in the 1930s for non-ferrous metals. Its application to steel took some time—the higher melting point of steel and its lower thermal conductivity presented technical problems that actually required several decades to overcome. Continuous slab casting came to dominance in the 1960s, and eliminated ingot steel-making and the rolling of steel from ingots. The savings in energy, labor and metal were significant. The next, related development—the thin-slab caster—was introduced by the American mini-mill steel-maker Nucor in 1989. It added a significant advantage to continuous casting, eliminating the "thickness reduction process" of slabs produced by integrated mills which, at 8 to 10 inches (22.3cm to 25.4cm) were four to five times thicker than the thin slabs. The new process not only saved time, energy and money but changed the relation between integrated steel-making and mini-mills by extending the latter's capability and thus altered a significant element of steel-making economics—not least in reducing the amount investment required to produce thin slabs.

11   In *New Steel's* review of twentieth-century steel-making two technologies are identified as being responsible for the greatest leaps in steel productivity: the basic oxygen furnace (BOF) of the late 1950s/early 60s (from its development in Austria by Voest-Alpine, whose first furnace became operative in 1952, and was commercially commissioned in 1953); and thin-slab casting which arrived in the late 1980s.

12   The seeds of EAF technology were first sewn by Humphrey Davy's discovery of the carbon arc in 1800. William Siemens brought this concept to furnace technology, and patented the electric arc principle in 1878. However, it was first commercially applied by Paul Heroult in France in 1886 to produce aluminium. Steel was made by this method in the following year, but not perfected until 1900. Small furnaces were in commercial use shortly afterward, but they were mostly used to make specialist alloys from re-melted selected alloy steel scrap. The technology was further advanced during the First World War when it was used in the making of high-grade alloy steels. Carbon steel was not produced until several decades later; however, it was not until the creation of mini-mills in the late 1950s/early 60s that the EAF came into its own (on this history see W. K. V. Gale "Origins and Development of Small-scale Steel-making" in R. D. Walker (ed.) *Small Scale Steel-making* Barking (UK): Applied Sciences Publishers Ltd., 1983, pp. 1–19). Over the past 40 years or so, continual increases in the improvements of EAF technology have meant that "tap-to-tap" times have becomes increasingly shorter (currently around 70 minutes, which is 30 to 40 minutes less than it was a decade earlier), while at the same time the weight of steel tapped continually increases. Equally, during the same period, electric-energy consumption fell from 450 to 390 kilowatt hours/ton. The key issue for EAF production is the availability of scrap steel. To counteract this problem "virgin iron" scrap substitutes, from processes like direct reduction, are being used

or developed as materials management mechanisms in relation to the economics of scrap, especially its market availability and cost (this either as a substitute when prices are high, or as an alternative when only a very limited supply of scrap is available). The scrap picture, of course, alters very significantly between old industrial and industrializing nations.

13 William T. Hogan "The changing shape of the Chinese steel industry" *Iron Age-New Steel* October (web edition, p. 4/10).

14 John Schriefer "Privatising steel in Latin America and Asia" *Iron Age-New Steel* July 1997 (web edition, p. 10/17).

15 See Tony Fry *A New Design Philosophy: An Introduction to Defuturing* Sydney: UNSW Press, 1999, which seeks to show, via historical and contemporary examples, how much of the industrially manufactured designed world has taken futures away (defutured), as well as providing the means to read this process—which it called defuturing.

PART THREE

TOWARDS THE AGES OF SUSTAINMENT

FIGURE 7.1   De-industrialization, Coventry, UK, 1980s. Photo by Tony Fry

7 ENVIRONMENTS OF IRON AND STEEL-MAKING

L et's begin with two images from two different times and places. The first is a nineteenth-century iron and steel works, let's say in Britain or perhaps the USA. The works are surrounded by a town which has grown rapidly in its wake, a dense agglomeration of factory buildings and cramped, homogeneous housing. There are no green spaces, no trees or any other kind of vegetation. Smoke billows continuously from tall factory chimneys and seen from a distance the town is permanently enveloped in a dark haze. Looking more closely, a fine film of dust, cinders and ash covers every surface, not just buildings and machinery, but clothes on washing lines, goods on display in stores; it is even ingrained into the very lines of the faces of the inhabitants. At midday, lights in the city are burning because visibility is so poor in the polluted atmosphere. The town's river becomes increasingly putrid as it flows past the steel works and other factories receiving their wastes. Railway lines cut right into the heart of the town, bringing trains carrying coal and ore to the steel works and adding further to the smoke and dust that is everywhere. Slag and other wastes are piled high and dominate the landscape. Inside the steel works is like a scene from hell with "terrible noises, shooting, thundering and lightning" and big trains carrying vessels of fire.[1] Men work in conditions of extreme heat, noise and danger, wheeling barrows of raw materials to feed furnaces and working within a few meters of molten iron or steel, their limbs wrapped in wet rags, as they go about their work of tending and tapping furnaces, casting, puddling, stacking hot "pigs" or loading pigs onto wagons, tending an earth-shaking, ear-splitting steam-driven forge hammer and innumerable other arduous jobs. Women and children can be seen in the yard doing work such as breaking up and sorting lumps of ore and coal. Adjacent to the main works, there might be chain-making shops operated by families from their homes.

By contrast, our early twenty-first century integrated steel works is located in a coastal town, some raw materials arriving by ship, others by rail. There are many houses not far from the works, but they are not covered in grime; instead gardens flourish, even within the grounds of the steel works itself. Gone are the thick black haze, the pervasive soot and the obvious signs of polluted waterways. Slag and other wastes are still piled high, but they are contained within an area screened by trees, and the piles grow and shrink according to the fluctuations in markets for these secondary materials. Besides being much cleaner, the operations themselves are dramatically quieter and eerily automated. Only a few maintenance workers can be seen in the actual works, other workers are in remote-control booths responding to electronic data on video screens or in laboratories analyzing samples of molten steel that have been delivered direct from the furnace by pneumatic tube. All the charging, smelting, pouring and other heavy processes occur in an almost depopulated space. There are also new types of workers: environmental managers, who monitor emissions, report to environmental authorities, liaise with local communities, record and analyze environmental performance data for the company's annual environmental report.

This chapter will tell the story of how the move was made from the first to the second image, but with qualifications, for, as we shall see, it is not a narrative of simple and unqualified progress. The two generic images are evoked intentionally to demonstrate difference.[2] They of course do not tell the whole story in several important respects. First, at any point in time different kinds of steel-making processes, working conditions and steel-making communities are in coexistence. "State of the art" and "leading edge" are exactly that: *atypical* of the moment. What might be thought of as nineteenth-century conditions continued, or have been newly established, in many parts of the world well into the twentieth century. Second is the coexistence of different methods of steel-making: the advantages of the less noisy and remote-controlled operations of a modern basic oxygen furnace do not apply to the deafening explosions of an electric arc furnace.

We will find that the improvements in environmental conditions of iron and steel-making and the reduction in impacts on biophysical environments that have occurred over the last hundred years have been driven primarily by economic factors rather than as a response to environmental imperatives or because steel-makers have become "good environmental citizens," as many companies like to claim. We will also find that while the kinds of environments created by steel-making are in many instances today much cleaner and less intrusive, invisible pollutants continue to be discharged into the atmosphere and into waterways, with effects that are complex and cumulative rather than obvious and immediate. More significantly, it will become apparent that it is only possible to claim improvement in environmental conditions if one is working with a restricted, biophysical model of "environment," which is preoccupied with measuring "impacts." Such a model, which is dominant within environmental science, forms the basis of environmental regulation and of corporate environmental "performance indicators." Its limitation is that it is not capable of recognizing or dealing with the impacts associated with the kinds of ecologies brought into being by steel (or any other industrial material) that were discussed in previous chapters: those forces of production, process, product application and use that have constituted and transformed the world as "modern" and which have inscribed so much of what is now perceived as unsustainable.

Our thinking needs to go beyond the frequently reductive empiricism of environmental science and the often rigid legalism of regulation that seeks to classify, quantify and set limits on "environmental releases" in order to control them. While this kind of activity does have value in curbing some of the more excessive polluting practices of industry (in fact we will present an historical assessment of it in relation to the steel industry in the following chapter), it fails to work with a picture that is sufficiently relational and shies away from the difficult issue of *structurally inscribed unsustainability*. To counter this, our approach will be to give an account of certain environments and ecologies of iron and steel-making that have existed at different times and places, attempting to weave together a discussion of biophysical impacts with other impacts less amenable to incorporation by environmental science. Our contention is that seeking to understand the *fundamental nature* of the processes of iron and steel-making, and the kinds of environments that they create, puts one in a much better position to define problems and to pose appropriate material and cultural solutions.

Our account will be structured around specific ecologies and sets of exchange relations within particular environments. The ecological model to be used does not depend upon a clear divide between the natural and the artificial, whereas this divide is fundamental to environmental science and to the environmental impact approach. We will see how particular material exchanges such as the extraction, transport and processing of ore and fuel create distinctive environments which in turn impact upon other environments. The processes are extremely complex, and can only be briefly described here. Our starting point is the localized ecologies of mining and fuel production for iron-making, then to describe the multi-directional impacts that occur as one fuel, i.e. coal, came to dominate. This will lead us to the environment of the industrial city. We will then return to consider the changing environments of iron and steel-making, adding flesh to the two skeleton images with which the chapter opened.

Localized ecologies: The environments of mining

The first stage in iron and steel-making is obviously obtaining the raw materials. The mining of ore (and later of coal) creates changes to the landscape and has localized effects on biophysical ecologies. This has been the case over many centuries wherever mining has occurred, and whatever the metal being sought.

The immediate effects of this activity include direct damage to vegetation from excavation and from transportation of the ore from the mine site. Even the illustrations in the sixteenth-century *De Re Metallica* show dead and damaged trees around mine sites. As vegetation dies, soil is exposed and becomes vulnerable to erosion. Shaft mining creates tunnels, which can create soil subsidence and alter ground water movement. Quarrying results in large depressions that, unless filled and re-vegetated, become permanent features of an eroding landscape often collecting run-off and turning into lifeless ponds. Historically, surrounding forests were often cut to obtain timber for mine props and roof supports, thus deforestation in mining areas, particularly in the vicinity of coal mines, was common.

Iron ore mining is generally less environmentally destructive than mining for coal or other ores such as copper, gold or silver that occur in much lower substrate concentrations. Iron ore is the most common of metals. As already observed, it makes up a significant percentage of the earth's crust. Although of varied quality, deposits are relatively plentiful. Iron ore yield is within the 40 to 60 percent range, which means lesser quantities of gangue (waste) result, as compared to copper mining where yields are generally less than 1 percent. Iron ore is usually obtained from open cut mining. While underground mining has less visual environmental impact, it can have dramatic consequences for ground and surface water movement. For example, pumping allows underground mining to extend below the water table, but when mining is completed, the pumps no longer operate and the pits fill with water to form lakes.[3] So, while this provides a visually acceptable result (as opposed to open scars on the landscape), what is established after mining is always a new ecology, it is never the same as it was before mining. But, excepting where mining has caused the eradication of a species because it was conducted in an area of last remaining habitat for particular species, the new mix of species that is established afterwards is not in itself a problem; the issue is the sustainability of what is put in place—i.e. that the plant colonies will be self-sustaining, compatible with adjacent areas and provide habitat for creatures.

Perhaps the most significant impacts derive from the essence of mining itself, which is the separation of valuable ore from unwanted material (gangue). Mined material brought to the surface is crushed and washed to assist separation. Further site operations for iron ore may include magnetic separation to reduce the amount of tailings transported to the smelter. Thus, large piles of crushed rock, sand and clay tailings accumulate around mine sites. These can collapse resulting in slides that can end up in a nearby river, entirely blocking its flow; there are even instances of slides that have obliterated nearby human settlements.[4] The cumulative impacts of mining are significant, but are not always immediately apparent: metallic sulfides in mineral or coal deposits once exposed to air and moisture form sulfuric acid which is transported to streams by artesian springs or by run-off from exposed tailings dumps; this can kill plant and animal life in receiving streams and rivers, with effects persisting for many years after mining has ceased; similarly trace heavy metals can be washed into nearby waterways causing fish kills; sediment deposition can reduce the size of river channels and increase downstream flooding, causing vegetation dieback, disruption to human settlements and agriculture.[5]

As well as creating a particular kind of environment with impacts extending well beyond the immediate site, the activity of mining creates a dangerous environment for those directly involved, the miners themselves. Mining always has been, and still is, physically dangerous; underground explosions and collapses causing multiple deaths are still not uncommon occurrences around the world. Historically, the worst prevailing conditions have pertained where the economic incentives of miners, or more typically of mine owners, have overridden all other concerns.

Robert Gordon, in his history of American iron-making, observes that coal mines have generally been located in valleys and out of sight, that nineteenth-century mine owners were investment-driven and miners were on quantity-based contracts. All of these circumstances created little incentive for safety, with the result that mines were woefully inadequate in terms of provision for ventilation, escape routes and roof supports. He notes that roof falls caused far more deaths than the more dramatic explosions and fires in mines that were reported in the newspapers of the day.[6] But by the beginning of the twentieth century, deaths due to explosions were on the increase. Between 1890 and 1906 a total of 22,840 coal miners were killed in US mines. Then, in 1907, there were some particularly shocking mine explosions: 500 miners were killed in two West Virginia mines and 75 miners were buried alive in Alabama; 3,200 in total were killed in similar accidents in that year. The US government's response in 1910 was to create the Bureau of Mines, giving it responsibility for improving mine safety. The Bureau took a technical approach, establishing mine safety and experimental stations throughout the country, which tested and licensed explosives and electrical equipment for use in mines and conducted investigations into mining accidents. Regulation and inspection of mines was left to the states until 1952 when annual inspections by federal inspectors were introduced.[7] But appalling safety conditions in coal mines is not confined to the past. China, today the world's largest coal (and steel) producer, has a disastrous safety record. In 2002 alone, nearly 7,000 miners died, and between 2000 and 2009 there were 30,000 accidents and more than 51,000 deaths.[8]

It would be incorrect to assume that mining environments have become safer due to some kind of evolutionary process or simply because of "progress." In some cases, pre-industrial conditions were less hazardous and had much lower environmental impacts, if only because of their smaller scale. Regulation and inspection also are not just features of recent history. In fact, in the eighteenth century Sweden had a highly regulated iron industry, with the government keeping detailed records on the production and sale of iron from every forge. Licenses were issued stipulating the quantity each establishment was permitted to produce, this being controlled through a College of Mines and district "courts of mines."[9] While historical evidence about the nature of mining environments is very uneven, occasional instances of difference can be glimpsed. Coxe, a British traveller who visited Sweden in 1790, gives an account of the Dannemora mine, worked since 1448 and famous for the quality of its ore and the iron produced from it. He described the miners' villages each with "three or four regular streets, often planted with trees, a church, a school and a hospital." Rather than being accessed via subterranean shafts, the ore body of the mine was reached by much wider abysses or gulfs, which had been created by excavation. Descending one of these to a depth of 500 feet in a large bucket operated by rope and pulley, Coxe felt giddy and terrified as the mine inspector who accompanied him sat on the edge of the bucket using a stick to touch the sides of the rock and steady their descent. Coxe observed with amazement three girls standing on the edge of the ascending bucket knitting "with as much unconcern as if they had been on *terra firma*."[10] What this anecdote reveals is that these mine workers had developed a mode of being and a culture in which danger was ever-present but able to be accommodated. It is also the same, as we will see, with the hazardous conditions of the iron and steel mill.

Agricola, the author of *De Re Metallica*, identified a number of miners' diseases from his experience as a physician from 1527 to 1530 in Joachimsthal, Bohemia, a metal mining town of several thousand people. He observed that the cold conditions of mines induced rheumatism; that dust caused breathing difficulties, with some dusts being corrosive and causing consumption. He wrote of a particular mine with a black powder which ate wounds and ulcers to the bone (probably zinc oxide) and of corrosive cadmia (probably arsenic cobalt) which ate away at the feet of workmen when they were wet and injured hands, lungs and eyes. He noted the bad air in mines caused by lighting fires in shafts and tunnels to heat and break rock faces, and from poisonous elements mixing with standing water, creating toxic fumes. He recommended that miners wear high boots of rawhide, gloves to elbows and loose veils over the face. Moreover, Agricola devoted the second half of Book 6 of *De Re Metallica* to advice on ventilation of mines, describing and illustrating wind-driven, horse- and human-powered devices for forcing fresh air down mine shafts.[11]

The accounts of conditions in mines in England in the nineteenth century, in which children as young as 8 were forced to work hauling coal for up to 16 hours a day, are part of the infamy of the Industrial Revolution. And while many children died as the result of accidents, it was chronic health problems of the kind identified three centuries earlier by Agricola that shortened lives and foreclosed on possibilities for a better life which was the typical legacy of that historical moment.[12]

As mining increased in scale and intensity, so too have its negative effects. In many parts of the world, the richest ore-bearing bodies had been depleted long before the industrial era. Thus, methods have been developed for mining metals from lower grade ores. For example, in 1900, it was not feasible to extract copper from ore with less than 3 percent metal content, but improved techniques now mean that it is possible to process ores with less than 0.5 percent yield. But in order to extract the same amount of metal (or more), larger volumes of raw material have to be milled, which means a greater number of, and/or larger mining sites and generally, larger-scale operations. This in turn means that more mining waste is produced.[13] In 1991 the average grade of ores was 40 percent for iron (i.e. 60 percent waste yield), 23 percent for aluminium (77 percent waste yield), 2.5 percent for nickel (97.5 percent waste yield) and just 0.91 percent for copper (over 99 percent is waste!).[14]

Because the kinds of environments produced by mining activity as described above, are, by now, very well known, such knowledge can be used prefiguratively to avoid or lessen negative effects. Therefore, today, best practice for new mining operations involves several years of study to understand the environmental and socioeconomic impacts over the life of the project and to determine the most effective ways of dealing with them. This includes designing measures for managing tailings, erosion, dust and run-off as well as for liaising with affected communities. Planning for post-mining rehabilitation, so as to restore the land to a healthy condition, is also part of the pre-mining process.

So far, we have spoken of ore and coal mining as if they were identical activities, which clearly they are not. Coal is implicated in a range of complex ecologies, which will be dealt with below. But before looking at these, there are some points to note about the ecology of charcoal production, the fuel that iron-makers depended upon before coal.

Localized ecologies: Forests and charcoal-making

We have already seen that felling trees to make charcoal for iron production was a cause of deforestation in pre-industrial northwest Europe and that measures (such as establishing plantations, coppicing and regulation of cutting rates) were introduced in an attempt to maintain supply. Conservation techniques had been part of European and Asian cultures for many centuries. We have dealt with these in ancient China, but it is also worth noting that Japan, since 1600, had laws regulating forest use and limiting timber consumption and by the late eighteenth century, nearly all of Japan's forests were under some form of regulation or management.[15] At the same time in North America, in contrast, the recent European settlers perceived forests as abundant and endless, energetically clearing them for lumber, agriculture and for the charcoal-making of ironmasters. While it is estimated that charcoal production accounted for only 1 percent of deforestation across the whole continent, local impacts were often significant. Robert Gordon's account of the American iron industry tells of eighteenth-century proprietors of the Union blast furnace in New Jersey who destroyed a forest of nearly 20,000 acres in less than 15 years and then abandoned their investment for want of wood. In America, it was relatively easy for an iron master to move to a new area of virgin forest where there were plenty of suitable trees to harvest, so there was little interest in re-afforestation.[16] Some proprietors of iron works did manage their woodlands to yield a continuous supply of fuel, and the charcoal ironworkers' association, which controlled large tracts of land by the 1880s, emphasized, in their journal and conference papers, the importance of forest conservation.[17] But by then, coal had overtaken charcoal as the iron-making fuel.

Charcoal-making created a distinctive local ecology which had a major influence upon the charcoal burner's familiar environment. Timber would be cut in winter and set aside to season. The charcoal was made by heating wood in a pit, kiln or beehive oven. The charcoal burners waited for the windless conditions of summer until they lit their fires. Once lit, the fires had to be watched and tended almost constantly for two weeks, as the wood progressed through the stage of "sweating" and then carbonized down through the pile. The charcoal burner read the color of the smoke of the wood pile, controlling the process through covering and uncovering. When the charcoal was ready, it was raked from the pit and separated into small piles which had to be observed closely as they had a tendency to ignite. Charcoal makers lived in huts close to where they worked; each would have from two to six helpers and at any one time they would be tending up to six pits or kilns.[18] Skilled colliers found that their skills were very much in demand in America, and they could earn good money, but although they worked in the open air, they dwelt in an environment of pervasive smoke, which must have adversely affected their health. When colliers did operate kilns near towns, residents complained of the pungency of the smoke and often put pressure on them to move away. One visitor to an Adirondacks iron-making community in the 1870s described thin columns of smoke from charcoal kilns on distant hills, the constant passage of huge charcoal vans and loads of iron ore past huts and cabins populated "with smutty faced children," charcoal dust filling "every chink and crevice" settling on trees "and when it rained the leaves shed rivulets of ink." Charcoal continued to be used for American iron-making up to the end of the nineteenth century. Later developments included making it in the type of retorts mentioned earlier which allowed byproducts such as methanol and acetate of lime to be collected. The resulting charcoal was much stronger and could be transported by rail in specially designed cars.[19] But this was to be a short-lived phenomenon as coke began to overtake charcoal as the iron-maker's fuel of choice.

Everywhere ecologies: Coal

The displacement of charcoal by coke did nothing to improve environmental health. The production of coke in fact created worse environments and more widely dispersed impacts. Iron-makers no longer needed to be located near woods; instead they could order coal from dealers which, unlike fragile charcoal, could be transported over long distances. They then converted coal into coke *in situ*. This was done first in open fires, then in individual beehive ovens and later, in batteries of continuously burning ovens. Coke is made by heating coal under controlled conditions to drive off the volatile substances, particularly sulfur. Early coke ovens released large clouds of smoke and steam (the latter produced when the coke is quenched at the end of the process) and where there were many ovens, ash fell constantly on surrounding areas and the air was rarely smoke-free. Sulfurous fumes from coke ovens damaged crops and vegetation and caused damaging chemical reaction with various building materials. Byproduct ovens were introduced (e.g. in Pennsylvania in 1895) which reduced emissions of smoke and sulfurous fumes, but they released heavy pollutant loads in waste water.[20]

More significantly, the shift from charcoal to coal changed the situation from one of contained local impacts to transportable impacts extending over large regions. This is where the ecology and economy of coal, iron and steel meet. This is also where we see a move from localized ecologies impacting on extant environments to a material and its manufacture being constitutive of whole ecologies and environments. This also means that accounting for impacts becomes far more complex and difficult to accommodate into the descriptive approach of the foregoing account. The relationship between coal and steel operated like a feedback mechanism that propelled industrial development onward, spatially across regions and quantitatively in terms of increased output of product.

We get the first hint of this at Coalbrookdale with two important developments: one was in 1767 when iron rails were made for trans-porting raw materials, which soon freed ironmasters from having to be located near their raw material, particularly fuel sources. Second was the fact that the new fuel and carbon source, coke (the use of which had been pioneered at Coalbrookdale by Abraham Darby in 1709) was structurally far more robust than charcoal, allowing for a much heavier charge to be taken in a furnace which in turn enabled the construction of larger furnaces. Larger furnaces required a more powerful blast, which was provided increasingly by the use of another recent invention, the steam engine (with its coal-fired boiler).[21]

By the end of the eighteenth century, the stage was being set for a massive increase in iron and steel production as well as a vast expansion in coal consumption. Thus the circumstances for rapid reduction in air quality and an unprecedented increase in atmospheric carbon dioxide emissions were being put in place. Retrospectively, we can now see a quantum leap in the anthropogenic contribution to global warming occurring at this time.[22] This was a synergistic process: coke-fired furnaces driven by coal-fuelled steam engines boosted iron and steel production; iron and steel were used to manufacture rails and rolling stock (in fact in the middle years of the nineteenth century the most prolific single iron item produced by iron puddling furnaces was railway lines)[23]; in turn, the chief commodity carried by railroads right up until the Second World War was coal.[24]

In Britain, Europe and the United States from roughly 1850 coal was the major fuel source for industry, transport and households until its displacement from around 1950 with oil and gas. Modes of moving coal had structural impacts and created new physical, social and economic environments. A major impetus for canal building in the UK in the eighteenth century and the USA in the nineteenth century was to move coal on barges. Canals made long linear alterations to landscapes, they cut across districts and reconfigured patterns of local exchange according to where bridges were placed. Canals altered local water supplies and drainage patterns and sometimes precipitated flooding.

Railroads similarly extended the structural impacts of iron-making, but even more so: deforestation due to timber cutting for sleepers; scoring and making deep cuttings into the landscape with the laying of lines; a new danger to animals from locomotives. Then there was the noise, smoke and cinders brought by trains as they sped through the countryside and into the centers of towns and cities.[25] A new kind of sprawling, grimy, lifeless environment was created in the form of railway sidings and yards. As Lewis Mumford put it: "from the 1830s on, the environment of the mine, once restricted to its original site, was universalised by the railroad."[26]

Everywhere ecologies: The industrial city

But more than this, the combination of coal, the railroad, steam power and vastly increased iron production was the basis of a new kind of everywhere environment: the industrial city. Charles Dickens dubbed this "Coketown," a motif taken up by Lewis Mumford, who described the process in which a city gets structured by the nature of industrial production: factories claimed the best waterfront sites because of their need for water in cooling and other processes; and waterways provided a convenient place for factories to dump their wastes which are carried downstream—damaging banks, vegetation and marine life along the way. Housing for workers was an afterthought, built rapidly, often on land filled in with ash and other industrial waste. Along with child labor, the living conditions of workers was the other infamous feature of the industrial city: housing was notoriously overcrowded, poorly ventilated, without piped water or plumbing. Arguments by sanitary reformers such as Edwin Chadwick that such conditions were the ultimate cause of the high levels of pauperism, disease and premature death among the British working class eventually persuaded authorities to put in place municipal sewerage works and regular refuse collections.

While industrial production increased in complexity and linked together distant regions (as suppliers of raw materials and fuels or as markets) into new sets of relations, at the same time a reduction in complexity, or what Mumford calls "unbuilding," was going on: craft modes of production, social structures of village communities, the public places and amenities of older towns were all being undone. So too was the diversity of plant and animal life as species were depleted or made extinct through habitat destruction and the degradation of soil and waterways.[27] A new regime of economic imperatives ruled that produced an environment "neither isolated in the country nor attached to a historic core ... (which) spread in a mass of relatively even density over scores and sometimes hundreds of square miles."[28] Mumford evokes the character of generic Coketown, whatever part of the world it was in, whether Sheffield, Birmingham, Pittsburgh or Lille, as an environment of incessant hammerings, with the clang of engines, roar of steam and hissing of water, where "black clouds of smoke rolled out of the factory chimneys ... soot and cinders everywhere," where the "oil and smudge of soft coal spat everywhere" and added to these:

> ... constant smudges on flesh and clothing, the finely divided particles of iron from the grinding and sharpening operations, the unused chlorine from the soda works, the clouds of acrid dust from the cement plant, the various by-products of other chemical industries: these things smarted the eyes, rasped the throat and lungs, lowered the general tone, even when they did not produce on contact any definite disease.[29]

It wasn't just factories and their use of coal that created this kind of environment. Coal became the major power source of urban civilization. It was used to heat homes and workplaces. It was used as the fuel of the boilers that powered trains, ships, ferries and elevators. The USA overtook Britain as the world's biggest coal producer in 1913, accounting for nearly

40 percent of world output (Britain was 22 percent). At this time more than 75 percent of the USA's energy was supplied by coal, with consumption skyrocketing from 20 million tons in 1860 to 650 million tons in the peak year of 1918. Coal was something every householder was familiar with. The well-off purchased it by the ton and stored it in coal cellars, the poor bought it weekly by the bucket. Different types, grades and lump sizes were available at different prices and for different kinds of furnaces or stoves. With this intensity of coal use, smoke, cinders, ash and coal dust became even more pervasive in the urban environment. The smokiness of the air was exacerbated by the type of coal used; most of the cities of the eastern states used bituminous coal which was more abundant and cheaper, but also had higher sulfur content and produced more smoke (New York was an exception, where the harder anthracite was used). Bituminous coal was also favored by the steel industry as it made better coke.[30]

Pittsburgh, a major center of steel production, was infamous for its poor air quality. It was the first American city in which soft (bituminous) coal was used widely by industry, and as early as 1823 one resident lamented: "the increased number of chimneys pouring forth dark and massive columns of smoke, begins to be felt as an almost intolerable nuisance."[31] By 1890 Pittsburgh had 21 blast furnaces, 49 iron foundries, 15,000 coke ovens and 33 rolling mills.[32] Photographs of downtown Pittsburgh in the 1920s and as late as 1945 show scenes of smoky darkness even at noon.[33] At this time it was estimated that Pittsburgh's air pollution meant that an additional $2.3 million annually was spent on general and domestic cleaning (this didn't include buildings).[34] The degraded environment extended well beyond central Pittsburgh, with major iron and steel works within a 100-mile radius in the towns of Homestead, McKeesport, Braddock, Donora, Monessen, Aliquippa, Youngtown, Altoona and Johnstown. And while Western Pennsylvania, centering on Pittsburgh, was the largest steel conurbation in nineteenth- to early twentieth-century America, there were also major steel works in South Chicago and Joliet in Illinois; Wheeling, Ohio; Lackawanna, New York; Gary, Indiana; Cleveland, Ohio and Birmingham, Alabama.

The ecology of iron and steel works

Iron and steel production was a catalyst for industrial development and a significant contributor to the degradation of urban environments, particularly of air quality, as will be discussed further in the following chapter. But what of the environment of the steel works itself? Clearly, it is difficult to generalize given the many changes that have occurred over time and across cultures. But some major shifts can be registered, and some already have in earlier chapters, such as the move from more craft-based, highly skilled, labor-intensive processes of iron works to the automated production of the integrated steel works. In what follows, these shifts will be evoked via some generic, fragmentary images: the first set of observations aims to give some sense of a mid-nineteenth-century iron works, the second will concern early to mid-twentieth-century steel-making. Both (re)constructions will be discussed as different kinds of environments of work, of space and of ecologies of materials.

The iron-making environment

As we have seen, the shift from bloomery to blast furnace in the West brought about a change in scale and intensity of iron-making. Where it was possible for one person to operate a bloomery, with the process able to stop and start at will, a blast furnace required at least a dozen persons and it needed to be operated continuously for several weeks or months. One would assume that the differences between the two processes (smelting ore with charcoal in a hearth to produce a bloom to be worked with a hammer versus smelting in a furnace to produce liquid iron which is then run out into molds to cool into pig iron) would be absolutely determinate. Establishing and running a blast furnace required substantial investment to construct the works, diversification of tasks and coordination of work processes. Yet bloomeries also achieved a high level of complexity in the United States where they operated up until the 1890s, while the bloomery process had been abandoned by British ironmasters a century earlier. Gordon describes Adirondack forge bloomers who worked their hearths continuously for six days a week on 12-hour shifts, each forge running several hearths operated by highly skilled bloomers working in coordination with hammer-men. The forge of S. P. Bowen in New York in 1870 employed eight bloomers working four hearths supplying blooms to two hammer-men; it also had a charcoal kiln and ore separator run by inside contractors and depended on 50 laborers to dig ore, cut and deliver wood to the charcoal maker and to haul the fuel and ore to the works.[35] According to Gordon, in America the bloomery process was pushed almost to the limit of efficiency possible within the constraints of its chemical process and that compared to the highest development of bloom smelting in Europe, the American bloomery used 24 percent less charcoal, 34 percent less ore and 86 percent less labor to produce the same quantity of iron.[36] This drive towards efficiency was also to be found in the emergent steel industry and it propelled America to overtake Britain as the world's biggest steel producer by the early twentieth century.

At a blast furnace there were two main teams: those doing the charging at the top of the furnace and those on the casting floor. The charging work was arduous, extremely hot and dangerous: ore, fuel and flux were loaded into wheelbarrows, weighed and then pushed across to the charging platform, the furnace keeper shut off the blast, the charging door was opened and the fillers tipped contents of the barrows into the furnace, ore fuel and flux alternately. They might be doing this every 15 minutes in 12-hour shifts. There was always the risk of the contents of the furnace shifting while the door was open, causing flames, smoke and dust to shoot out and engulf them. Conditions in the casting house were no better. This is where liquid slag flowed out and was directed into slag pits, and where the molten iron was tapped from the furnace and channeled into sand casting beds. Before a tap, the casting crew formed the wet sand into long channels (runners or "the sow") and smaller side channels ("pigs") taking care not to add too much water, for if they did, this would cause a steam explosion when the molten iron made contact. Wearing wooden clogs to protect their feet, the casting crew worked within a few feet of the white-hot liquid iron, controlling its flow into the pigs; then they shoveled sand over the runners to keep them hot while the pigs cooled which made it easier to break them free, which they did using iron bars and sledge hammers. Further heavy, and no doubt often injurious, work followed after the pigs (each weighing 100 pounds) had cooled sufficiently to be lifted by hand and stacked outside the casting house.[37]

Particular conditions were required for establishing a blast furnace: transport for raw materials and finished product, either by rail or a navigable waterway; space to store raw materials; roasting ovens for pre-treating the ore prior to charging; easy access to the top of the furnace into which ore, fuel and flux were fed, which meant building the furnace next to a steep bank or providing elevators to lift it to a charging platform; power was needed to create the blast, which initially meant water power from a watercourse and later steam power to drive the bellows of a blowing engine. Also needed were reliable supplies of water for cooling processes and space for storing or a means of removing process waste, especially slag. What was needed in short was a *system*, a lot of space in the right location and a workforce for handling, moving and processing materials. Initially this was conceived of as an open system of inputs and outputs, with little concern for the effects of these as they impinged on environments beyond the works.

But process improvements during the nineteenth century, particularly those focusing on greater efficiency, gradually created closed loops, such as the reuse of waste heat. As we saw in Chapter 4, Scottish engineer James Neilson demonstrated in 1828 that by pre-heating the air used for the blast, fuel could be saved and a more stable furnace environment created. Given that a blast furnace produces combustible gases in vast quantities, it made sense to use it in the process itself. In 1845, J. P. Budd of Ystalyfera, South Wales patented a method for using blast furnace gas to heat stoves and boilers,[38] and by 1854 Andrew Ure described a number of methods of furnace gas pre-heating in his *Dictionary of Arts, Manufacturing and Mines*.[39] Refinements followed, culminating in the regenerative furnace which had brick-filled chambers in which waste furnace gas was burnt, with air for the blast then being pre-heated by passing it over the bricks.[40]

As we know, iron smelted in a blast furnace requires further refining to be workable. In the late eighteenth century and in much of the nineteenth century it was re-smelted and worked in a puddling furnace, or, increasingly, it was converted to steel. The puddling process, as patented by Henry Cort in 1784, separated the fuel source from the metal in a "reverberatory furnace," one advantage being that coal could be used, because, not being in contact with the metal, it wouldn't be able to contaminate it with sulfur. Another advantage of the process was that the metal could be worked while in its molten state, the aim being to drive off the impurities with heat and end up with a ball of white-hot iron which was then removed from the furnace for further working by a steam hammer or put through rollers to form iron bars.

As has already been noted in Chapter 4, puddling was the heaviest form of regularly undertaken labor there has ever been, requiring extraordinary judgment and skill as well as physical strength, involving, as it did, the working of balls of molten iron of up to 200lbs in weight. Puddling was performed in conditions of extreme heat, glare and smoke.[41] It was very energy-intensive, with heat for the furnaces being supplied by coal; this was particularly polluting when the high sulfur content bituminous coal was used; also the nature of the process was such that it often produced a smoky flame. As is the case with the production of pig iron, puddling also produced large amounts of slag which were dumped on nearby open spaces, accumulating into large piles. The slag still contained a significant amount of iron, which blast furnace proprietors sometimes collected to add to their charges.[42]

Given what is now known about the nature of emissions from iron-making and their effects on human health, it can be assumed that blast furnace workers, puddlers and virtually all other kinds of iron and steel workers suffered many acute and chronic conditions due to their continuous exposure to high levels of tarry smoke, dust, carbon monoxide, sulfur dioxide, nitrogen dioxide, hydrogen sulfide and volatile organic compounds. The prevalence of "black lung" disease (silicosis) among coal miners is well known, and those handling coal at iron works would have been just as vulnerable. Other respiratory diseases caused by constant exposure to toxic dust and fumes would have been prevalent. Then there were other ever-present risks: burns from hot liquid and solid metal; impact injuries; eye damage from grit and glare; acid burns from pickling liquor. Workers who constantly handled hot iron were said to develop hands that resembled reptile skin from the build-up of scar tissue from repeated burns. And who knows what other chronic conditions were induced by the constant dehydration from the extreme heat? Certainly the conditions dramatically affected men's appearance—many a 30-year-old looked 50. One employer, interviewed for a US federal investigation into labor conditions in the iron and steel industry, commented that for the task of top filling a blast furnace where work was performed in temperatures up to 128 degrees, "gorilla men are what we need."[43]

Iron and steel-making in the first and second industrial ages was both replete with danger and environmentally hellish. What does contemporary steel-making look like?

The ecology of modern steel-making

An integrated steel works is usually a vast sprawling environment in which complex exchanges of materials and energy take place. It is a product of industrial evolution and it has come to take on a distinctive ecological character—it is an ecology in which excess is always produced. In some instances the integrated steel works approaches, but has never quite achieved, the industrial ecologist's goal of total closed loop production.

The use of waste from one process to feed into another within a steel mill encompasses heat, gases and solid residues. These process developments have been economically driven; this is particularly so in terms of the heat economy. The following presents a summary of these material exchanges and that which is excessive to them—unwanted pollutants.[44]

Whether an integrated steel works or an electric arc furnace, pollutants are produced at every stage of steel-making—either airborne, water-borne or solid waste. Today it is coke ovens and sinter plants, as well as the overall high level of CO_2 emissions, which are deemed to be the major environmental problems for the steel industry. Before considering these in more detail, it needs to be kept in mind that the generation of polluting residues is an intrinsic part of *all* industrial production across all its stages. This is because of the fundamentally unnatural nature of the processes involved: the extraction, concentration and transportation of desired substances, their transformation via heat and chemical reactions to produce new compounds, the release of emissions to air and water, and the disposal of unwanted solid matter. Moreover, this movement and transformation of "the raw materials of nature" has occurred in greater volumes over the last 200 years.

Certain stages in industrial process carry particularly high risks of direct or indirect, immediate or long-term damage to living systems. The more obvious, highly visible forms of pollution have been recognized and sought to be managed for a long time. This is the case for example with the losses and dust emissions from handling and transporting bulk raw materials. Air- and water-borne pollutants pose more of a problem as they are sought to be treated via progressive phases of dilution, separation, concentration and/or containment. Frequently the form of treatment itself creates a further pollution problem—for example, the wet scrubbing of gases significantly reduces smoke stack emissions, but it also generates polluted water which then requires treatment before further use or release. The treatment of polluted water in settling tanks then creates the problem of having to dispose of toxic sludge. Dispersal of pollutants into the natural environment—air, soil, rivers, oceans—where by dilution and biological processes they will eventually be rendered harmless, is no longer an option because the sheer quantity of industrial pollutants now well exceeds the "sink capacity" of the environment. Also it is now known that many of the toxic byproducts of industry, such as heavy metals, have a tendency to bio-accumulate, causing chronic damage over an extended period rather than immediate symptoms.

Heat reuse

As we have seen, one of the first "waste products" to be reused was heat. The regenerative principle of using blast furnace gas to provide the heat to maintain the process was the first step in heat recycling, but there are other variations on the use of blast furnace exhaust gas, such as using it in co-generation technology or to fire coke ovens. In turn coke oven gas may be used as supplementary heat for the blast furnace or it may be used to generate electric power for the rolling mills, or even to feed into the regional grid.[45] Clearly, major heat savings were achieved once the practice of charging molten pig iron directly to the converter became widespread (instead of having to re-smelt iron pigs and work them in a puddling furnace). Equally, contemporary hot slab casting technology that avoids having to re-smelt steel before it can be cast is also more energy-efficient.

Water cycle

Huge volumes of water cycle through a steel mill. For instance, coal is washed prior to coke-making, the finished coke is quenched, water is used for the "wet scrubbing" of exhaust stacks, cast steel is cooled with water before going to the rolling mill; water is also used in slag crushing operations and for cooling and cleaning throughout rolling and finishing, including the de-scaling of rolling installations and rinsing in the pickling process. One calculation is that the manufacture of a tonne of steel uses between 8,000 and 12,000 liters of water. Another is between 100 and 300 cubic meters per ton of crude steel, but up to 90 percent of this may be recycled within the plant.[46] This requires that water is treated before moving from one process use to another. The recycling of water in all these operations is driven both by economic and environmental imperatives.

The finishing processes for steel generate large quantities of polluted water. Mill scale, rust, oxides, oil, grease and dirt are chemically removed from the steel prior application of protective coatings. One way this is done is through hot acid treatment (usually hydrochloric or sulfuric acid) known as "pickling," which is followed by rinsing. Alkaline cleaners such as caustic soda, soda ash and phosphates may also be used to remove mineral oils, animal fats and oils from the steel surface prior to cold rolling. Spent pickle liquor is sometimes passed through a recovery unit, returning clean acid to the work baths and generating marketable byproducts like ferrous sulfate and ferric oxide; caustic soda can also be recovered. After use and extraction of byproducts, process water is eventually returned to the environment, either through the public wastewater system or into natural watercourses. Its environmental effects depend upon the degree of treatment it has received before release. But it's not just a matter of removal of suspended particulates and dissolved chemicals, for the temperature and quantities of discharged water can also be damaging to waterways and their life forms. Excessive quantities of oxygen-seeking effluents discharged into waterways will compete with aquatic micro-organisms that require oxygen for life. A higher temperature of discharge waters to receiving waters can also encourage the growth of certain micro-organisms at the expense of others and have knock-on effects up the food chain and alter the ecology of the waterway.

The main water-borne pollutants arising from the steel industry are solid particles in suspension, hydrocarbons, acid wastes, oils; and cyanides, thiocyanides and phenols from coke-making.

Air pollution

The main air pollutants produced in steel-making are tarry smoke, dust, carbon monoxide, oxides of nitrogen, oxides of sulfur particularly sulfur dioxide, hydrogen sulfide, fluorine compounds, volatile organic compounds (VOCs) and, as has been discovered more recently, dioxins. Large volumes of carbon dioxide are produced in iron and steel-making from the burning of coal and coke. While not a pollutant as such, CO_2 is the principal greenhouse gas responsible for global warming. Clearly, airborne pollutants are a threat to workers in the immediate environment of the steel works, to adjacent communities and beyond.

Coke batteries and the sinter plant are often the largest structures in an integrated steel works, as well as the most troubling sources of pollution.

Pollutants from coke-making

A coke oven battery comprises a series of ten to 100 individual ovens, side by side, with a heating flue between each pair. The process of converting coal to coke produces numerous unwanted substances driven off in the form of gas, which can be recaptured and converted into chemical compounds for other industrial processes. These include naphthalene and tar which find end uses in products such as plastics, paints, wood preservatives and electrodes; ammonium sulfate which is used as fertilizer; and BTX (benzene, tolulene and xylene) which are on-sold for the production of styrene monomer, a base ingredient for polystyrene plastics.

Coke-making is a concern in terms of air emissions of coal dust, CO, CO_2, HS, SO_2 and benzene. Many of the air- and water-borne pollutants from coke-making are "scheduled" hazardous substances, the release of which is legally required to be recorded and reported to regulatory bodies on a regular basis. Even though pollutant loads are lessened because of the recovery of the previously mentioned byproducts, toxic residues remain in the waste water used to extract these. The quenching of coke produces large quantities of water contaminated with VOCs, phenols and suspected carcinogenic particles. More polluted water accumulates from moisture in the coking coal and from condensation of steam used in charging the coke ovens. Water-borne pollutants are dealt with in some plants by the use of biological effluent treatment plants to remove cyanide, phenols, ammonia and hydrocarbons from process water. Air pollutants have been reduced in some more advanced plants by using pulverized coal injection which substitutes coal for coke in the blast furnace, thus reducing the amount of coke needed by between 25 and 40 percent. An advanced method of eliminating VOCs and suspected carcinogenic particles is the use of dry quenching. Historically, one of the biggest problems with coke ovens has been gas leakage from poorly sealed doors.

Pollutants from sintering

The reuse of blast furnace gas is partly responsible for the introduction of another form of in-house recycling via the sinter process. Although the principle of sintering, which is the fusing together of granular materials into a homogeneous mass, had been known for some time and is used in other metal industries, its take-up in the steel industry dates from about 1910 when it was introduced as a way of managing flue dust extracted from blast furnace exhaust gases.[47] Cleaning of gases is necessary because the pressure of the blast causes numerous particles of coke, ore, lime, etc., to be driven off, and, if blast furnace gas is to be reused, these particles have to be removed otherwise they will wear out the gas engines through abrasion. This is done by causing coarser particles to be deposited directly in precipitators (dust catchers), trapping medium-size ones via a series of baffles, and capturing finer particles by passing through washing towers. These methods have all been in use since the 1910s.[48]

The sinter process developed to incorporate a variety of process wastes. As well as dust from the blast furnace and from the steel converter, the sinter plant also takes in raw materials fines (fine ore, coke breeze, fine fluxes) and mill scale (small flakes of iron oxide from rolling mills), which would be too light to add directly to the blast furnace. All these materials are mixed, moistened and spread in a layer on a looped conveyer where gas burners ignite coke, the heat causing all the particles to agglomerate. Thus the sinter cake gets "baked" as it successively passes down through the horizontal layers of the plant. It is then broken up, cooled, screened and travels on conveyers to the blast furnace.

A sinter plant is responsible for a significant share of total emissions from a steel plant, particularly atmospheric releases of SO_2 and dust from handling, storage and blending of fines. Dust can be captured, but the SO_2 is very difficult to eliminate.

More recently it has been discovered that sinter plants emit dioxins. These are a broad family of complex chemical compounds which do not exist naturally, but are formed in combustion processes where hydrocarbons and chlorine are present. Dioxins have been found to be carcinogenic, to have disruptive effects on animal hormones and the capacity to bio-accumulate, which is to say concentrate up the food chain, for example by being stored in animal fats.

Dioxins exist in very low concentrations and can only be detected with sophisticated equipment, but they are extraordinarily potent and pervasive; they are a particularly interesting example of the contemporary problems and controversies associated with defining environmental impacts and appropriate action in response to them. This issue will be taken up in the next chapter, as will the question of how the impacts of other pollutants generated by steel-making are sought to be minimized.

While steel-making, if not adequately controlled and regulated, can impact negatively on biophysical ecologies, a steel works is also a key nodal point in a local or regional industrial ecology, which if well managed can lessen the negative biophysical impacts. Exchanges of waste materials occur between a steel mill and other industrial producers: we have seen that the byproducts of coke-making and the extracts of spent pickling liquor are used as raw materials in industries such as plastics and chemicals. But in terms of volume, the two most significant areas of material recycling in the steel industry are the export of slag for other uses and the importation of scrap metal as feedstock.

Slag reuse

Large volumes of slag are produced by blast furnaces, and since the eighteenth century uses for slag have been sought, such as molding it into blocks or using it for cement, the latter first being considered in 1871. In its composition, slag is a cementitious material with a high lime content that is very close to cement.

The most common use of blast furnace slag is as an aggregate in concrete and as road base, though other uses are possible, such as using it as a soil conditioner. It is made available for such uses either by being left to solidify then crushed, or the liquid slag can be converted to granular form by running it past water jets which can be controlled to produce the desired grain size.[49] The disadvantage of this process is that it drives off large amounts of hydrogen sulfide and sulfur dioxide. Slag from steel-making (BOFs and EAFs) requires stabilization before being used as aggregate in road construction, this to counter its tendency to expand.

At an integrated steel works, blast furnace slag is likely to make up over 30 percent of solid waste and slag from steel-making about 20 percent, the former having a wider range of uses. As a ground cementitious material, granulated iron blast furnace slag can be used as a substitute for ordinary Portland cement, although a small percentage of cement is still needed to accelerate curing. The significance of this, besides substituting a recycled for a virgin material, is that the use of slag cement can reduce the greenhouse gas contribution of concrete, the world's most prolifically used construction material. Besides the carbon dioxide emissions from the energy used to manufacture cement, significant amounts of CO_2[48] are released by the nature of the process itself, specifically the *calcination process* (i.e. when the calcium carbonate in limestone is changed into calcium oxide). Cement makes up 10 to 15 percent of the average concrete mix and the cement industry accounts for 8 to 10 percent of worldwide greenhouse gas emissions, second only to fossil fuels. Within an industrial ecology, the use of slag cement therefore has definite environmental benefits.[50]

Slag from iron-making has a higher reuse rate than steel slag. It is claimed that nearly all blast furnace slag in the USA is reused, while only about half of the slag from EAFs and BOFs is reused (some as cement substitute), with the balance being landfilled.[51] It should also be noted that the uptake of slag has often been obstructed by the cement industry.

Recycling scrap

Iron and steel, as materials, are inherently and endlessly recyclable and the recycling of scrap metals is almost as old as the working of metal itself, with smiths frequently forging discarded iron into new forms. Surplus armaments have provided a prolific source of scrap iron: Henry Cort's first experience of iron-making was at a Hampshire forge which was mainly dedicated to recycling old iron for the navy.[52] At the Harpers Ferry Armory in Virginia the recycling of borings and trimmings was an integral part of production, with about a quarter of the gun iron produced in 1819 being derived from recycled scrap.[53] At the end of the American Civil War, the Tredegar Iron Works in Richmond, Virginia, started purchasing old cannons and worn-out railroad equipment as raw material, combining it with new puddled iron. The pace of technological change, much of which was driven by innovations within the iron and steel industry, fed the "scrap stream" as machinery increasingly became obsolescent, the substitution of iron rails by steel being particularly significant. The Tredegar Works even built a roll train for breaking down old rails and by 1875 was purchasing 20 carloads of scrap wrought iron a week, much of it coming from old bridge parts and railroads which were being rebuilt in steel.[54] Even the steel industry's own equipment is grist for the scrap mill. English steel-masters visiting some of Carnegie's plants in 1900 were amazed that equipment still perfectly functional could be scrapped just because an improved version had become available.[55] There is also a long tradition of recycling scrap from within the production system. Steel companies with both integrated works and EAFs often feed scrap from the former to the latter.

As has been noted, one of the advantages of the open-hearth method of steel-making was that the furnace could take a full charge of scrap. And while this was not the case with the Bessemer converter, later developments of it, specifically the basic oxygen furnace, had the ability to take scrap as part of their charge. The ultimate development has been the electric arc furnace which was virtually designed to run on a diet of scrap; it can only be charged with scrap or with suitable "scrap substitutes," dominantly DRI which comes in a variety of forms.

At the end of the twentieth century, when environmentalism turned recycling into almost a moral crusade, an industry such as steel, which was utilizing waste material to such an extent that it regarded virgin material as a "substitute," would have been regarded as very advanced. Certainly the steel industry has played this card, with corporate environmental information constantly stressing the high rates of steel recycling compared to other materials like plastics. But such a view can only be maintained if one ignores that scrap recycling has always been a part of the iron and steel industry, and that in some instances it has actually assisted in accelerating production (and all the pollution associated with it). At the very least, the recyclability of steel needs to be seen as a double-edged sword. Yes, it means savings on the mining of new ore, but also the ease of its recyclability accelerates obsolescence. If proof of this is needed, one only has to consider whether the rate of turnover of car models and styles and thus the volume of production of the car industry would be as intense as it is, if not for the convenient recyclability of steel car bodies. Now imagine the same rate of production, but with car bodies made of a material like PVC, which is very difficult to recycle and has a lifespan of about 400 years. There would simply not be the landfill space available to deal with such a situation.

Scrap is a legitimate raw material, not a second-best substitute for pig iron; in fact, because it has already gone through at least one refining process, in some cases it is superior (e.g. when it is not contaminated with traces of other materials like oils or plastics). It is not surprising then that the buying and selling of scrap metals became an industry in its own right. The overall rate of steel recycling in the USA is claimed to be 65 percent and for appliances as high as 81 percent.[56] As we have seen, scrap is a more prolific source material for steel-makers in old industrial, rather than newly industrializing nations. It is estimated that in the past 50 years more than 50 percent of the steel produced in the USA has been recycled.[57]

Of course, the quality of scrap varied. Low-grade "dirty" scrap (steel with oil, paint or plastic coating) increased air quality problems, not least because of the higher levels of dioxins produced when such material is combusted.

Notes

1 According to one European immigrant worker's description of his first experience of an American steel works. Quoted in David Brody *Steelworkers in America: The Non-Union Era* Cambridge, MA: Harvard University Press, 1960, p. 100.

2 The nineteenth-century picture draws on descriptions found in Robert B. Gordon *American Iron 1607–1900* Baltimore, MD: The Johns Hopkins University Press, 1996, Lewis Mumford *The Culture of Cities* London: Secker & Warburg, 1938 and Charles Dickens' evocation of "Coketown" in *Hard Times* of 1854. The contemporary description is based partly on the BlueScope (formerly BHP) Steelworks at Port Kembla, Australia.

3 Gordon op. cit., pp. 27–31.

4 An example of each: on October 22, 1997 a slope failure at BHP Copper's Pinto Valley (USA) mine released 230,000 cubic meters of mine rock and tailings into the bed of Pint Creek (a removal and rehabilitation program followed); on October 12, 1966 a flow slide of colliery waste buried part of the Welsh coal mining village of Aberfan, killing 144 people. *BHP Environment Report* Melbourne: The Broken Hill Proprietary Company Ltd, 1998, p. 31 and Frank N. Magill (ed.) *Great Events from History II: Ecology and the Environment Series* Volume 1, Pasadena, CA: Salem Press, 1995, pp 873–7.

5 Information from Gordon op. cit., pp. 32–3, 49; T. E. Graedel and B. R. Allenby *Industrial Ecology* Englewood Cliffs, NJ: Prentice Hall, 1995, p. 211 and BHP Environment Report op. cit., pp. 31–5.

6 Gordon op. cit., pp. 46–7.

7 Magill op. cit., pp. 80–3.

8 Guiling Wei "Statistical Analysis of Sino-US Coal Mining Industry Accidents" *International Journal of Business Administration* Volume 2, No. 2, May 2011, p. 82.

9 Harry Scrivenor *History of the Iron Trade from the Earliest Records to the Present Period* London: Longman, Brown, Green and Longmans, 1854, pp. 146–51.

10 The account is quoted in Scrivenor op. cit., pp. 147–8.

11 Bern Dibner *Agricola on Metals* Norwalk, CT: Burnly Library, 1958, 23, 66–71. *De Re Metallica* was completed by Agricola in 1556.

12 There were rare, heroic exceptions. An Open University text book on industrialization and culture contains two telling documents in this regard. The first is an extract of evidence given before a Royal Commission on Trades Unions on April 28, 1868 by Alexander MacDonald who began his life as a miner at the age of 8 in a Scottish ironstone mine, going on to work in coal mines. He described how he rose at 2 p.m. to travel to the mine where he and other boys worked in leather harnesses hauling loads until 5 or 6 p.m.; he also stated that of the 20-odd boys he worked with, he believed that he was the only one still living. MacDonald was 40 years of age when he gave this evidence. The document that follows his testimony is an extract from Boase's *Modern English Biography* published early in the twentieth century, giving an account of Alexander MacDonald's life. We learn that he saved money and put himself through Glasgow University, became a teacher, held office in the Scottish and National Mining Unions, agitated for 20 years against women and children working in mines and became the first Working Men's Member of Parliament. *Industrialisation and Culture 1830–1914* edited by Christopher Harvie, Graham Martin and Aaron Scharf, London: Macmillan/Open University Press, 1970, pp. 109–12.

13   Gary Gardner and Payal Sampat *Mind over Matter: Recasting the Role of Materials in Our Lives* Worldwatch Paper 144, Washington, DC: Worldwatch Institute, December 1998, p. 9.

14   Graedel and Allenby op. cit., p. 234. The extraordinarily low yield from copper mining means that management of tailings is crucial to a successful operation. This is apparent from the history of the Ok Tedi copper mine in Papua New Guinea where landslides in 1983–4 destroyed initial work on a tailings dam. This, along with the factors of high rainfall, steep terrain and geological instability led to the company to conclude that it was too risky to go ahead with the dam. After further environmental investigations it was decided to release the tailing into the Ok Tedi River. But by the mid-1990s sediments were building up in the river causing flooding, destruction of village produce gardens and reduction of the river's fish stocks. After villagers took BHP to court, in a case that extended over a number of years, an out-of-court settlement was reached with BHP agreeing to pay 150 million Australian dollars in compensation as well as several hundred million dollars' worth of remediation work (involving extensive dredging of the river). By mid-2000 remediation work had not gotten beyond the trial stage and BHP was seeking to close the mine. *BHP at Work in the Environment* Melbourne: The Broken Hill Proprietary Company Limited, January 1994, p. 15 and *BHP Environment Report 1998*, Melbourne: The Broken Hill Proprietary Company Limited, December 1998, pp 33–5 and Carolyn Batt, "PNG villagers issue writ" *Sydney Morning Herald* April 12, 2000.

15   I. G. Simmons *Environmental History: A Concise Introduction* Oxford: Blackwell, 1993, p. 152.

16   Gordon, op. cit., pp. 41–3.

17   Ibid., pp. 41–3.

18   Ibid., pp. 34–7.

19   Ibid., pp. 38–41.

20 Ibid., pp. 50–2.

21 J. R. Harris *The British Iron and Steel Industry 1700–1850* London: Macmillan Education, 1988, pp 34–6.

22 It should be remembered that greenhouse gases, especially carbon dioxide, can have a very long atmospheric life (200 years plus).

23 Norman J. G. Pounds *The Geography of Iron and Steel* London: Hutchinson, 1966, p. 21.

24 Lewis Mumford *The Culture of Cities* London: Secker & Warburg, 1938/45, p. 159. The economic significance of coal to some of the Eastern American railroad companies (51 percent of Pennsylvania Railroad's traffic and 56 percent of Baltimore and Ohio's) was such that they bought land in the coalfields to ensure supplies and profits. David Stradling *Smokestacks and Progressives: Environmentalists, Engineers and Air Quality in America, 1881–1951* Baltimore, MD: The Johns Hopkins University Press, 1999, p. 8 and 195 note 6.

25 Gordon op. cit., pp. 45–6.

26 Mumford op. cit., p. 150.

27 Ibid., p. 151.

28 Ibid., p. 191.

29 Ibid., pp. 194, 191, 192.

30 Stradling op. cit., pp. 7ff., 13, 12.

31 Ibid., pp. 37–8.

32 Bernard A. Weisberger *The Age of Steel and Steam Volume 7 1877–1890 The Life History of the US* New York: Time Inc., 1964, p. 49.

33 See Stradling op. cit., pp. 162, 170, 171.

34 Mumford op. cit., p. 193.

35 Gordon op. cit., pp. 95–7.

36 Ibid., p. 99.

37 Ibid., pp. 119–24.

38 J. W. Hall *The Metallurgy of Steel: Volume 11—Mechanical Treatment* London: Charles Griffin & Company Limited, 1918, p. 691.

39 Ure *Dictionary of Arts, Manufacturing and Mines*, fourth edition, pp. 1108–11.

40 Gordon op. cit., pp. 109–11, 310.

41 Pound op. cit. London: Hutchinson and Co., 1959, p. 20.

42 Gordon op. cit., pp. 148–9.

43 U.S. Bureau of Labor *Report on Conditions of Employment in the Iron and Steel Industry* (4 Volumes, Washington 1911–13), p. 299, quoted in David Brody op. cit. Cambridge, MA: Harvard University Press, 1960, p. 33. The high levels of industrial accidents in the many plants of the United States Steel Corporation began to attract negative publicity in the early 1900s, leading in 1908 to the establishment of a Safety Committee within the corporation with staff dedicated to workplace safety education. Serious accidents were reduced by 43 percent within four years. Gerald G. Eggert *Steelmasters and Labor Reform 1886-1923* Pittsburgh, PA: University of Pittsburgh Press, 1981, pp. 43-4.

44   Except where indicated otherwise the sources for the following account are: EPA Office of Compliance Sector Notebook Project, *Profile of the Iron and Steel Industry* Washington, DC: US Environment Protection Agency, September 1995; Economic Commission for Europe *Problems of Air and Water Pollution Arising in the Iron and Steel Industry* New York: United Nations, 1970; and BHP *The Making of Iron and Steel* Melbourne: BHP Steel Public Affairs Dept, 8th edition January 1998.

45   Pounds op. cit., pp. 26-7. An early example of the use of waste heat to generate electricity for external use was in the Middlesbrough district, where in 1908 the Cleveland and Durham Electric Power Company was purchasing coke oven gas, blast furnace gas and exhaust steam from local blast furnace operators. J. W. Hall *The Metallurgy of Steel: Volume 11— Mechanical Treatment* London: Charles Griffin & Company Limited, 1918, p. 715.

46   The first figure is from Simmons op. cit., p. 143 and the second from Economic Commission for Europe op. cit., p. 7.

47   BHP op. cit., p. 3.

48   They are described in J. W. Hall *The Metallurgy of Steel: Volume 11— Mechanical Treatment* London: Charles Griffin & Company Limited, 1918, pp. 691-2.

49   Gordon op. cit., pp. 168-9.

50   *Green Building Digest* 1995.

51   John Schreifer "Reaping the value from dust and slag," *New Steel* February 1997, web version, 2 of 9.

52   Harris *The British Iron and Steel Industry 1700-1850*, pp. 39-40.

53   The economic importance of scrap recycling was greatly emphasized in this government works, the chief of ordnance putting arguments to Congress about materials cost savings if funds would be provided for a forge dedicated to working up scrap iron into bars. Congress duly provided $5,000 in 1836 for this. Gordon op. cit., p. 193.

54   Ibid., p. 194.

55   David Brody op. cit., p. x. In 1937, railroads still remained the biggest source of scrap, a *Fortune* article reporting this with the telling comment, "every minute the iron and steel viscera of mechanical civilization is being outmoded. When it gets obsolete, it is scrapped. When it is scrapped, it is sold for regeneration into new metal." "Scrap is an Issue" *Fortune* May, 1937, p. 119.

56 The figure is for 1997. *The Appliance Recycler* (magazine: web version www.recycle-steel.org [downloaded April 1999]).

57 Ibid.

FIGURE 8.1   Pittsburgh steel mill, 1906. Carnegie Library of Pittsburgh

8 REGULATING INDUSTRIAL ENVIRONMENTS

An examination of the environmental regulation of the steel industry reveals the limitations and contradictions of state control of environmental matters, and in this sense provides a way of viewing the relation between the state and industry more broadly. Furthermore, such an examination can provide insights into something more troubling—which is the fundamental limits to the possibility of "sustainability" existing within the dominant economic and political structures.

The context of environmental regulation

It may seem contradictory that it is in prosperous, free-market economies that environmental regulation of industry is most developed, rather than in centralized planned economies. While China and the nations of the former Soviet Union have been notorious for lax environmental regulation, in the USA, Western Europe and Japan there are whole rafts of legislation that set limits on how industry handles its raw materials, wastes and emissions to air and water. Companies are compelled by the state to spend money on anti-pollution programs and equipment, this representing, from a strictly capitalist perspective, unproductive investment. The explanation of course lies in the democratic structure of these nation states, where governments are compelled to listen to citizens' concerns about environmental degradation, while "balancing" the demands of business to operate freely (and to provide jobs for citizens). This has set in train a reactive dynamic, in which regulations formulated as responses to highly particularized issues gradually accumulate and are eventually gathered under the umbrella of a state agency which is given responsibility for "the environment." This, for example, was the history of the USA's Environmental Protection Agency. While a thorough understanding of this process as it has unfolded over the last century would require an analysis of the nature of bureaucracy and the nation state, there are some observations that can be made. For instance, the state must appear to be acting in rational and fair manner. An activity does not become regulated unless harm can be proven, and here, the authority of empirical science meets the authority of the legal system: science defines environmental problems and laws are framed accordingly. Particular substances are designated as toxic, science is deployed to determine the effects of their release into the environment, their impacts upon various life forms and ecosystems; then acceptable levels of release are set, maybe along with a program for progressive reduction agreed upon by negotiation between the state and the industries concerned. Monitoring, testing, data gathering, regular reporting to state environmental agencies are all part of the apparatus of environmental regulation, as are methods of proof and attribution. In fact, the empirical study of environmental problems has mushroomed as it draws on diverse technical specialisms such as toxicology, epidemiology, risk assessment and the full range of biological and chemical sciences.

The nature of impacts

Over the twentieth century, the pressure for environmental regulation has been driven by concerns about human health and safety as well as concerns for the condition of "nature."

The concept of "environmental impacts" has become central to the attempts to reduce the damage of industrial development. The meanings of "environment" and "impact" that are assumed in this taken-for-granted concept are very problematic. First, "environment" is posited purely in biophysical terms: as habitats, wilderness, as the space of undisturbed nature and as separate from human, especially urban, activity. An impact is considered to be anything that harms the assumed natural condition of stability. Much of the pressure for environmental reform is inspired by the desire to protect idealized environments that in fact no longer exist or are selected "endangered" areas of special status. A retroactive, even nostalgic environmentalist world view has underpinned much environmental reform over the last one hundred years. This is apparent, even today, in the photographic images of pristine nature used by environmental groups to publicize their causes. But the same kinds of images are also increasingly seen in the annual environmental reports of large companies.

Environmental regulation has been most developed in Western democratic countries, where governments have been compelled to respond to public concerns about pollution, but it is also within their political systems that it is layered onto fundamental, irreconcilable differences. For if an environmentalist ideology based on the idea of autonomous nature has been one driver of reform, coming from the opposite direction is a belief in the inherent rightness and necessity of industrially driven economic development. In the democratic context, what results from the struggle between the two positions is compromise: governments seek to curb the excesses of industry by imposing limits on pollutants permitted to be discharged into the environment, while industry complies in order to demonstrate responsibility and good environmental citizenship to both its market and the wider community. What has not been sufficiently recognized is that negative impacts are structurally part of the increasing scale and intensification of human beings' appropriation and transformation of the materials of "raw nature." Thus the disjuncture between the extremity of environmental problems and the political actions taken to address them ever grows. This is seen most starkly with global warming: where greenhouse gas output actually needs to be cut by between 40 and 60 percent, yet it has been a struggle to get agreement, via the Kyoto Protocol, of reduction targets for developed nations of a mere 5.2 percent.[1]

The point being made is that the limited effectiveness of environmental reforms stems largely from the failure to confront head-on the contradictions between the desire to conserve "natural environments" and the desire to pursue industrial development. Environmental impacts get classified in terms of emissions of certain toxic substances to air, water and soil; limits on these get set, then production processes are modified to be less polluting.

Over the twentieth century, knowledge has increased significantly about the complexity and inter-relatedness of environmental impacts, while their control and regulation has followed a different pattern—one of accumulation of fragmented, highly specific legislation. As we will see in the following case study on smoke and air pollution, environmental problems have come to be seen as increasingly complex, and at the same time the science of classifying and measuring them has become more elaborate. As this has happened, the authority of science has grown, as the arbiter of "safe" levels of emissions, acceptable concentrations and other parameters, which then become inscribed into regulations. While it is clearly useful for pollutants to be understood with greater precision, there is a downside: measurement and classification can become divorced from a relational understanding of industrial production, and furthermore, impacts which are not amenable to empirical calculation frequently do not get taken into consideration at all. Impacts do not always reveal themselves immediately or in normal circumstances—our picture of them is limited. Therefore, the emissions that regulators seek to control have to be recognized as figures within a restricted (world) picture of "environmental impacts." The regulatory approach is not able to take into account that which has been emphasized in foregoing chapters: that steel (and the coal upon which it depends) has immeasurably transformed the world; that its production, the technological innovation associated with it and the uses to which it has been put, have been major drivers of the character of the modern world.

Air pollution case study 1: "The smoke problem"

The history of pollution control in the steel industry is inseparable from the history of naming and seeking to deal with the emergent environmental problems of the modern industrial city, especially air pollution. While individual steel works were (and still are) sometimes identified as significant point sources of pollution, it has been the quality of the air over whole cities that manifested as a problem, one created by the combined fuel usage of industrial, commercial, domestic and transport.

Although there are some isolated historical examples of the regulation of air pollution, including a London ordinance of 1306 limiting the burning of coal, systematic regulation did not occur until very late in the nineteenth century, with US cities taking the lead.[2] Not surprisingly, it was in US east coast cities like Pittsburgh that air pollution first emerged as a political issue, with citizens forming organizations to agitate for reform from the 1890s onwards.

The focus was very much on symptoms: the issue around which citizens first agitated was "smoke prevention," characterizing the problems of urban smoke variously as cutting out sunlight and thus retarding plant growth, causing buildings to deteriorate, increasing laundry bills, creating general discomfort and contributing to premature deaths.

Chicago adopted a smoke ordinance in 1881 that fined industries for emitting dense smoke; similar ordinances followed in other industrial cities: Cincinnati, Cleveland, Detroit, Pittsburgh, Salt Lake City, St Louis and St Paul.[3] Based on the English common law principle of the right of municipalities to regulate "nuisances," smoke ordinances were often vague and not able to be upheld when tested in court.[4] Complainants had to prove that emissions from a particular source were excessive and causing them ill-effects, when the actual problem was more typically the poor air quality of the urban environment which was the result of emissions from many thousands of large and small point sources, including domestic ones.

The first wave of anti-smoke reformers were middle-class women who formed groups and campaigned against urban smoke using arguments about its detrimental effects on cleanliness, health and aesthetics. They sometimes called on medical expertise to support them, but medical opinion was divided as to whether urban air pollution in itself constituted a significant health hazard. One of these groups, the Health Protective Association of Pittsburgh appeared to have won a victory when in 1892 the city council passed a law prohibiting any chimney or smokestack from a stationery boiler to emit bituminous coal smoke. But the ineffectiveness of this point-source approach, as well as a more fundamental reluctance to challenge industry, was demonstrated in the implementation of this law: the fact it applied only to stationery boilers meant that locomotive emissions were excluded, but more significantly it applied only to a certain district, the boundaries of which were drawn by the city council specifically to exempt iron and steel works and other heavy industry.[5]

Such action (or rather creative inaction!) was not exclusive to Pittsburgh, PA: in many other US industrial cities, the anti-smoke ordinances that were on statute books were never enforced (e.g. in Gary, Indiana, where United States Steel was the largest and most influential company) or they were watered down to accommodate industry's interests. In Birmingham, Alabama where the steel industry dominated the local economy (half of the city's wage earners worked in steel foundries, mills and machine shops, while thousands more worked the county's iron ore and coal mines), businessmen pressured to have a 1912 anti-smoke ordinance repealed. Arguing that the ordinance would force them out of business, they succeeded in gaining major amendments that were so lenient that the city's three smoke inspectors all resigned, claiming that the new law left them with nothing to do. Then in 1915, Birmingham's major manufacturers successfully lobbied the state government to remove municipal authority over air pollution.[6]

The other tendency in the American history of air pollution control was the prioritization of technical solutions. Municipal smoke inspection which had initially been the responsibility of city health departments became the province of engineers; soon it was mandatory for all smoke inspectors to have engineering qualifications. Through the Smoke Prevention Association, formed by municipal smoke inspectors in 1906, the smoke problem was defined as solvable through engineering knowledge. Leading members of the Association who were mainly mechanical engineers trained in coal burning technology believed that coal would remain the major industrial and commercial fuel and that it could be made to burn without smoke by the application of correct techniques, such as furnace design and adjustment, selection of coal type, the method and rate of feeding coal into furnaces. They paid little attention to cleaner fuels that could substitute for coal, such as gas. The Association promoted and disseminated information on the clean burning of coal; it worked towards setting standards for the operation of furnaces and boilers; it encouraged a cooperative relationship with industry, mobilizing arguments about cost and efficiency, rather than public health. Smoke inspectors argued that black smoke coming out of stacks was unburnt fuel and therefore represented a waste of money for the factory owner.[7] The problem could be solved either by properly operating existing equipment, which might also involve educating the operators, or by installing new equipment such as the Murphy Automatic Smokeless Furnace manufactured by the Murphy Iron Works of Pittsburgh, which gained the respect of engineers because it could demonstrate that its own furnace burned with no smoke.[8] Not surprisingly, the steel city of Pittsburgh became a center for the development of pollution control equipment. Members of the Smoke Prevention Association at their 1913 conference were taken on tours of Pittsburgh plants such as the Carnegie Steel Company's Homestead Plant to see smoke abatement appliances in action. By this time the Association had extended its membership beyond government-employed smoke inspectors to include anyone with expertise in smoke abatement, thus opening the door for industry to become members.[9]

The first comprehensive study of the smoke problem was undertaken between 1911 and 1914 by Pittsburgh University's Institute of Industrial Research (later renamed the Mellon Institute). It was multidisciplinary, the research team including physicians, architects, engineers, chemists, an economist, psychologist, bacteriologist, botanist, meteorologists and others. The findings, which were published in nine volumes and disseminated through trade and popular journals, focused on the effects of smoke on human health and psychology, on vegetation, weather, building materials and economic costs. Despite this comprehensiveness, only one volume addressed solutions. Significantly, this was titled "Some Engineering Phases of Pittsburgh's Smoke Problem."[10] This faith in engineering was not confined to the issue of urban smoke. The industrial city was virtually the product of engineers, especially of those who had invented and refined the major production processes, which had so accelerated the development of industry and the growth of towns and cities. American engineers, driven by the economizing directives imposed by steel-makers like Carnegie, had made major advances in the mechanization of steel production, facilitating increased output and thus increased air pollution.[11] At the same time, engineers were also called upon to solve the problems of the industrial environment, such as public health which was addressed by the provision of infrastructure for water supply, drainage and sewerage. Engineers made cities safer and more efficient by standardizing roads to suit motor vehicles and by designing mass transit systems. But where these kinds of extensive engineering infrastructure projects could have dramatic effects because they were establishing the infrastructure of cities, the engineering solutions proffered for the smoke problem, i.e. the retrofitting of furnaces and boilers, only happened when proprietors were convinced. Smoke inspectors in many US cities worked tirelessly in promoting and educating about ways to burn coal without smoke, yet overall air quality did not necessarily improve. There was also a weakness in the economic argument: saving money by burning coal more efficiently had little purchase in boom times or when fuel was cheap.

Ultimately, US industrial cities got cleaner air not because of improved methods of burning coal but because of the displacement of coal by cleaner burning fuels such as gas, oil and electricity (which although produced by coal-fired power stations, localized the smoke problem, and increasingly took it outside the boundaries of cities). Railways began replacing steam with diesel or electric power, this leading to smoke-free downtown areas, the railway companies often benefiting by developing the land above the electrified tracks (which would have been unthinkable in the steam era).[12] Laws were introduced in the 1940s in St Louis and Pittsburgh requiring all fuel users to install the more efficient mechanical stokers or to use smokeless fuels, with St Louis city authorities going so far as to purchase bulk supplies of cleaner coal for resale at competitive prices. These reforms were underpinned by the final recognition that the dominance of bituminous coal was the problem; at the same time inexpensive alternatives were becoming available, in particular, gas supplies.[13] In Pittsburgh, households heated with natural gas jumped from 17 percent in 1941 to 66 percent ten years later.[14] And by 1950 coal provided less than 30 percent of America's energy, whereas 40 years earlier it had been 80 percent.[15]

The history of smoke prevention shows us how a highly visible, but narrowly defined (and not fully understood) problem was incrementally addressed through a combination of public pressure, some regulation, but mainly through technological and economic means. But just as industrial cities were becoming less smoky, it began to be more widely recognized that air pollution was a complex phenomenon, not able to be reduced to smoke alone. An air inversion event in the steel-making Pennsylvanian town of Donora in 1948 brought this fact home to US regulators, while for lawmakers in the United Kingdom a smog incident in London in 1952 which killed 4,000 people was the turning point.

From the smoke problem to air quality

London had been famous for centuries for its "pea soup" fogs, due to its long history of coal burning. By 1200, Londoners had exhausted the supply of nearby wood fuel and had turned to high sulfur bituminous sea coal, and as early as 1306 an ordinance limiting the burning of coal was introduced.[16] Thermal inversion effects were described by John Evelyn in 1661, who spoke of smoke, fog and "the coldness of the air hindering the ascent of smoke," this when London's population was only 500,000.[17] By the late nineteenth century, as in the USA, there had been anti-smoke groups and some half-hearted attempts at regulation. But an event in 1952 was crucial. For four days in early December of that year, there were cold, windless conditions. This combination of low temperatures and still air meant that smoke could not rise vertically, so it just kept on accumulating, reducing visibility to zero and contaminating the air people breathed. Four thousand people died within two weeks. The incident sparked political outrage, changing the fortunes of a government. Three and a half years later a Clean Air Act was passed which provided limits on use of bituminous coal and subsidies for households converting from coal to electricity.[18]

In the incident in Donora, a suburb 25 miles south of Pittsburgh, it was industrial fumes, not coal smoke that were directly responsible for deaths. In October 1948 temperature inversion and fog trapped pollutants, mainly from the zinc works of the American Steel and Wire Company, creating heavy smog. After four days, 20 people were dead and nearly 6,000 were ill with symptoms of gasping and chest pains, hospitals being filled to capacity.[19] Then in 1950, the major component of Los Angeles' smog was discovered to be ozone created through the photochemical reaction of hydrocarbon exhaust and unburnt gasoline.[20] By now, public concern was shifting from "smoke" to "air pollution." Invisible gases rather than visible coal smoke become the new hazards as the economy switched from coal to gasoline.

While the understanding of air pollution was becoming more complex, it was sought to be regulated in highly specific ways, dealing with problems concerning specific pollutants as they arose, usually by setting limits on "end of pipe" emissions. In the USA, legislative activity became more intense in the latter part of the twentieth century: between 1895 and 1960 eight federal environmental laws were enacted, while between 1960 and 1990 over 30 were passed.[21] The initial assumption of regulators was that it was only localized areas that were affected by point-source pollutants. This led to "dilute and disperse" approaches to dealing with both air and water pollution, such as higher chimney stacks and longer pipes to release pollutants into air and ocean "sinks."[22] Such solutions often created new problems elsewhere, as with the example of acid rain, the source of which can be smelting, coal burning, automobile exhaust or even nitrogen fertilizer in agriculture hundreds of miles from where the rain eventually falls. While it had been known for some time that the burning of coal acidified rain, it was not till the 1950s that "acid rain" was recognized and not till 1966 in Sweden that a link between acid rain and decreasing fish populations was made. This raised the issue of air pollution across national boundaries, as it was industry in other parts of Europe causing acid rain on Swedish lakes. Ironically it was an anti-pollution technology—very tall stacks to carry emissions away from immediate vicinities—that contributed to acid rain by dispersing acid-forming compounds over wide areas (mostly sulfuric and nitric acid).[23]

Global warming

If acid rain is one dramatic manifestation of the interconnectedness of environmental impacts, the effects of increased carbon dioxide emissions demonstrate even more complex relations between industrial ecologies and other ecologies, with the impacts ultimately being global. The principle of CO_2-driven global warming was first proposed in 1896 by Swedish chemist Svante Arrhenius, who postulated that a doubling of CO_2 might raise average temperatures of the earth by 9°F. He made calculations indicating that the millions of tons of CO_2 being released into the atmosphere (as an effect of expanding industrialization and urbanization) could cause changes to the way heat was absorbed by the atmosphere, which would lead to increased temperatures and changes to the balance of life on earth. The significance of CO_2 to global warming is that it is transparent to incoming solar radiation but absorbs the infrared rays of the spectrum. Heat generated on earth is radiated as infrared rays which are absorbed by CO_2, which acts like a blanket, trapping heat in the atmosphere. Arrhenius's work was taken further by George S. Callendar, a British steam technologist and coal engineer, who studied the circulation of CO_2 in the earth, seas and atmosphere and examined long-term temperature records from 200 meteorological stations worldwide. He found evidence of increased temperatures and concluded this was attributable to increased CO_2 from the burning of fossil fuels. Callendar wasn't taken much notice of because most scientists then believed that oceans could absorb the excess CO_2 emissions, but later research established that oceans only absorb 15 percent, 40 percent stays in the atmosphere for decades up to 400 years, while 45 percent is absorbed by plants and soil.[24]

The balance of scientific opinion today is that human-generated CO_2 emissions are the cause of global warming, leading to significant climate change. In January 2001, the Intergovernmental Panel on Climate Change (IPCC) released its Third Assessment Report, which projected average global surface temperature over the period 1990 to 2100 rising from between 1.4 to 5.8°C. The imprecision of this prediction is not because of inadequate climate modelling, but rather because the level of greenhouse gases emissions (and thus the extent of the temperature increase) is determined by human action: at one extreme doing nothing about the problem and at the other, making dramatic reductions.

Successive IPCC reports have expressed increasing confidence that human activity is the main driver of climate change. Here are some statements from The Fifth Assessment Report released in September 2013:

Warming of the climate system is unequivocal, and since the 1950s, many of the observed changes are unprecedented over decades to millennia. The atmosphere and ocean have warmed, the amounts of snow and ice have diminished, sea level has risen, and the concentrations of greenhouse gases have increased.[25]

The atmospheric concentrations of carbon dioxide (CO2), methane, and nitrous oxide have increased to levels unprecedented in at least the last 800,000 years. CO2 concentrations have increased by 40% since pre-industrial times, primarily from fossil fuel emissions and secondarily from net land use change emissions. The ocean has absorbed about 30% of the emitted anthropogenic carbon dioxide, causing ocean acidification.[26]

Human influence has been detected in warming of the atmosphere and the ocean, in changes in the global water cycle, in reductions in snow and ice, in global mean sea level rise, and in changes in some climate extremes. This evidence for human influence has grown since AR4. It is extremely likely that human influence has been the dominant cause of the observed warming since the mid-20th century.[27]

CO_2 emissions are structurally at the heart of modern culture, whether we're talking about industrial processes or electricity from coal-fed utility companies to power heating, cooling, lighting, computers and the like in homes and businesses across the industrialized world. The effects of carbon dioxide emissions (and other greenhouse gases such as methane) are global and seeking to reduce them is a highly politicized and fraught international effort that exceeds the scope, imagination and will of current political systems.

Other atmospheric gases present in lower concentrations than CO_2, such as nitrous oxide and the chlorofluorocarbons, also absorb heat and contribute to the greenhouse effect. But the impact of chlorofluorocarbons (CFCs) on the earth's ozone layer is more significant. Invented in 1928, CFCs were widely used in aerosols, foams, refrigeration, air conditioners, solvents and fire extinguishers. In the early 1970s, it was discovered that CFCs have a very long lifetime and were reaching the ozone layer in the stratosphere (10–50km above ground) where they acted as a catalyst causing ozone (O_3) molecules to be broken apart. The ozone layer absorbs ultra-violet (B) radiation, limiting the amount reaching the earth's surface, which is important as an excess of UV-B causes skin cancers, suppresses immune systems, exacerbates eye disorders as well as having other harmful effects on plants and animals. Depletion of the ozone layer (especially any extension of the "ozone hole" over the Polar Regions) increases the incidence of these disorders. Under the United Nations Environment Program, a series of treaties were developed to phase out the use of substances that deplete the ozone layer, most significantly the Montreal Protocol of 1987, which has been amended five times with increasingly stringent requirements and has been ratified by 165 nations.

As scientific knowledge about the nature of the earth's atmosphere, of the dynamics of climate and the complex environmental effects of industrially produced chemicals has expanded, the case for caution and regulation becomes stronger, but also increasingly difficult—pollutants don't respect national boundaries, yet international action is very difficult to secure. We will now consider how the steel industry is environmentally regulated.

Control and regulation: The US EPA model

In the USA, anti-pollution laws began by *not* being industry-specific, but structured around that which was sought to be protected—air quality, water quality, public safety, etc. But increasingly industry-specific requirements have been added to Acts. This has culminated in the concept of "sector based environmental protection" with a greater emphasis on industry consultation and stakeholder involvement.

The US Clean Air Act has a number of amendments pertaining to specific steel industry activities, regulating for example:

- benzene emissions from coke byproduct plants;

- halogenated solvent cleaners used by the steel industry;
- chromium emissions from industrial process cooling towers;
- opacity and particulates in gases discharged from electric arc furnaces, argon-oxygen decarburization vessels and basic oxygen furnaces.

Besides emission standards, there are also requirements for continuous monitoring of designated air pollutants, and in the case of any new plants, the mandated use of the best available control technology. In 1991, 15 percent of the capital expenditure of the US steel industry was dedicated to environmental control, with most of this (80 percent) spent on air control measures. Most of this, in turn, was to keep coke ovens complying with the Clean Air Act.

Under the US Clean Water Act there are effluent limitations (based on the size of each facility) for 12 steel industry manufacturing processes: coke-making; sintering; iron-making; steel-making; vacuum degassing; continuous casting; hot forming; salt bath de-scaling; acid pickling; cold forming; alkaline cleaning and hot coating.

Significance of the steel industry as a polluter

We have already registered the gross material impacts of steel production in the previous chapter and given some indication of the production of wastes and of releases to air, water and land. The classification and quantification of these has become more elaborate as "pollution science" has developed.

A large part of environmental regulation involves the monitoring of industry's emissions and wastes. Laws concerned with air pollution, clean water, waste disposal and the handling of hazardous substances all have various provisions within them requiring industry to keep and submit records on procedures, incidents, the movement of materials, waste disposal and emissions. One such example is the US Environment Protection Agency's Toxic Release Inventory which requires companies over a particular size to report on discharges of some 600 chemicals and chemical compounds.

The categories used for gathering and presenting data are determined more by issues of legality than ecology. "Land releases" refers to the disposal of toxic chemicals within a facility's own boundaries (even though over time these might migrate beyond those boundaries). "Off-site transfers" includes disposal of toxic substances via sewers and landfill as well as to recycling and treatment facilities. As we saw in the previous chapter, the steel industry has well-established practices of byproduct recovery for use in its own processes as well as transferring wastes (such as spent pickle liquor) to off-site recyclers.

Regulatory bodies are compelled to deal with clearly defined industry sectors, and thus the data with which they work is not able to accommodate a complex ecological picture. So, for example, metal mining, primary metals and chemicals are all treated as separate industries, even though they exist in complex webs of interdependence: mining and primary metals all supply feedstock (as raw materials or byproducts) to the chemical industry, as well as being users of its outputs. Some of the steel industry's major toxic releases to land, air or water (by weight) are chemicals used in metal treatment processes: zinc and zinc compounds (steel-making in BOF and galvanizing of steel); chromium compounds (plating of steel); hydrochloric acid (pickling to remove scale from steel; trichloroethane (cleaning steel prior to coating); nickel compounds (steel-making). Other toxic releases of significance monitored by the US EPA include: nitrate compounds generated by blast furnaces; manganese compounds from steel-making; ammonia from coke-making; and lead compounds from BOF and EAF steel-making.

In the category of EPA classified "priority air pollutants" (which are classified separately from the Toxic Release Inventory), the iron and steel industry is the largest industry emitter of carbon monoxide (produced in coke-making, and also emitted from BOF and EAF furnaces). Iron and steel production is also among the top five industry emitters of NO_2, SO_2, particulates and micro-particulates (those of 10 microns or less). As we have seen NO_2 is associated with steel-making, SO_2 with iron-making and sintering, while particulates are released in iron-making, coke quenching, from BOF as oxides of iron or from EAF as metal dust with iron particulates, zinc and other materials associated with scrap.

The steel industry is just one economic sector among many (such as power generation, the automobile, building and construction industries) that is under pressure to reduce greenhouse emissions. The steel industry is the largest industrial energy consumer, accounting for about 4 percent of world consumption. It has high CO_2 emissions because of the need for carbon in the reduction processes of blast furnace iron-making, smelting and direct reduction processes—and in the case of EAFs because of the use of large amounts of electricity (where this coal-generated).

Air pollution case study 2: Dioxins

We have seen that the emission of CFCs and increased levels of CO_2 are accumulatively altering the nature of the earth's atmosphere. We have also argued that such fundamental problems go to the very heart of industrial culture, and that they cannot be solved by a single industry or government, but require concerted global action, which, if effective, will significantly change not only how industry operates, but the nature and distribution of consumer good and services, and thus lifestyles. Again, these issues are explored more fully in the following chapter. We would like to conclude this chapter's account of industrial pollutants by considering another group of chemicals—dioxins—that are having complex negative effects on a wide variety of living organisms, and that have, in fact, become deeply embedded in the structure of living matter across the globe.

Dioxins are yet another previously unknown hazard to come to attention in recent years. They are produced inadvertently under certain natural conditions (such as forest fires) and via industrial activities where chlorine is present (even in very small quantities). The term "dioxin" refers to some 30 compounds that have similar chemical characteristics and biological effects. Dioxins are very potent toxins that have been shown to interfere with normal growth and development in fish, birds, reptiles, amphibians and mammals.

Combustion (where hydrocarbons are present with chlorine), chlorine bleaching of pulp and paper, certain types of chemical manufacturing and other industrial processes can all create small quantities of dioxins. The two largest industrial categories of dioxin releases to the environment have been incineration of municipal and medical waste. Cement kilns that burn hazardous waste are also significant emitters, while the steel industry is probably third or fourth in order of significance.[28] But the biological effects of dioxins are more diffuse and complex than from direct point-source exposure. Dioxins tend to bio-accumulate, that is they concentrate up food chains, becoming stored in animal fats. Except for workplace exposure, 95 percent of human exposure to dioxins comes from consuming animal products (meat, dairy foods).[29] The potency of dioxins is illustrated by the fact that *total* annual emissions are measured in kilograms per year. Similarly, "tolerable intake" is measured in picograms (one trillionth of a gram) with the World Health Organization recommending a Provisional Tolerable Monthly intake of 70 picograms per kilo of body weight per month.[30] During recent decades, federal and state governments in the USA have introduced regulations concerning the operation of municipal and medical waste incinerators, resulting in emissions from these sources being significantly reduced

Within the steel industry, sinter plants and EAFs have been identified as sources of dioxin emissions.

What both processes have in common is the use of secondary materials, some of which may contain chlorine.

There are three theories about how dioxins come to be released in sintering: the presence of trace elements of chlorine in iron fines and coke breeze; the presence of precursor elements, i.e. other organo-chlorine chains which become altered in the sinter process; and "denovo synthesis," in which dioxins get formed at particular temperatures, destroyed then reformed.

While the processes of dioxin formation are not yet entirely under-stood, there are a number of emission control technologies that can reduce them, as well as process improvements that avoid or lessen their formation. Two examples of the former are: the Voest-Alpine AIRFINE process to control emissions from sinter plant using fine water mist scrubbing, in operation in plants in Austria, the Netherlands and Australia; and "carbon packed bed technology," a process which passes sinter plant waste gas through char filters that are later trans-ferred to a regenerator where they are heated at elevated temperatures to decompose and destroy the dioxins. This char filter process, which also removes particles, heavy metals, sulfur oxides and nitrogen oxides, is being used at a number of integrated steel facilities. An example of a process improvement that results in reduced dioxin, sulfur oxides and particulate emissions is the FASTMET process, an alternative to sintering, which converts iron oxide pellet feed, oxide fines and steel mill wastes into direct-reduced iron.[31] Activated carbon powder can also be used to reduce dioxin emissions from EAF steel-making (it is injected into the flue gas before it enters to baghouse). EAFs also achieve dioxin emission reductions when using direct-reduced iron instead of scrap as feedstock.

Clearly, steel production facilities are very costly entities, and improved processes or pollution control technologies are only introduced when it is economic to do so. Nevertheless, steel and other heavy industrial production facilities are under increasing pressure from governments to reduce harmful emissions. While frequently it is not possible to make a direct connection between, say, incidences of cancer in a particular community and pollution produced by a nearby factory, regulators of established industrial nations will seek to reduce risk. This is Environment Canada's stated policy on dioxins:

> Because dioxins, furans and polychlorinated biphenyls are highly persistent, bio-accumulative and toxic, continued release of these chemicals into the environment could unnecessarily prolong exposures, with a resultant increase in the risk to the environment and human health. Therefore, the goal of the Federal Government with respect to these substances is the virtual elimination of anthropogenic releases to the environment.[32]

The statement is linked to a program of reduction targets and reporting requirements for Canadian steel facilities (and other industrial sources) that are known emitters of dioxins.

It is to such global and futural concerns that our final chapter now turns.

Notes

1   The Kyoto Protocol, an instrument of the United Nations Convention on Climate Change, came into effect on February 16, 2005, the outcome of more than a decade of negotiation. Its overall target is to reduce carbon dioxide emissions to 5.2 percent below 1990 levels between 2008 and 2010. Because of variations in targets set for Ireland, Norway and Australia, and because Russia and the Ukraine were only required to stabilize emissions to 1990 levels, the actual targets for the USA, EU and Japan were set at 7, 8 and 6 percent respectively. The USA has refused to ratify the protocol; Australia, even though permitted to *increase* its emissions by 8 percent of 1990 levels, has also refused to sign. One of the main reasons the USA gives for not signing is that targets have not been set for "developing" nations, a position supported by Australia; the two biggest such emitters being India and China (whose emissions are one-seventh of the USA's).

2   T. E. Graedel and B. R. Allenby *Industrial Ecology* Englewood Cliffs, NJ: Prentice Hall, 1995, p. 78.

3   Frank N. Magill (ed.) *Great Events from History II: Ecology and the Environment Series, Volume 1, 1902–1944*, Pasadena, CA: Salem Press p. 44.

4   David Stradling *Smokestacks and Progressives: Environmentalists, Engineers and Air Quality in America, 1881–1951* Baltimore, MD: The Johns Hopkins University Press, 1999 pp. 61–2. The following account is greatly indebted to this very comprehensive historical study.

5   Stradling op. cit., p. 43.

6   Ibid., pp. 131–7 and p. 237 note.

7   Ibid., pp. 104–6.

8   Ibid., p. 91.

9   Magill op. cit., p. 45.

10 Stradling op. cit., pp. 98–100.

11 David Brody *Steelworkers in America: The Non-Union Era* Cambridge, MA: Harvard University Press, 1960, pp 1–7.

12 Stradling op. cit., pp. 157–8.

13 Ibid., pp. 164–71.

14 Ibid., p. 171.

15 Ibid., p. 186.

16 Magill op. cit., p. 549 and Graedel and Allenby op. cit., p. 78.

17 Magill op. cit., p. 549.

18 Ibid., pp. 550–3.

19 Ibid., p. 599.

20 Stradling op. cit., pp. 186–9.

21 Graedel and Allenby op. cit., p. 79.

22 I. G. Simmons, *Environmental History: A Concise Introduction* Oxford: Blackwell, 1993, pp. 146–7.

23 Magill op. cit., pp. 675–9.

24 Ibid., pp. 359–63.

25 Working Group I, Contribution to the IPCC Fifth Assessment Report Climate Change 2013: The Physical Science Basis Summary for Policymakers, September 27, 2013, p. 3.

26 Ibid, p. 7.

27 Ibid, p. 12.

28 Worldwide, the major source of dioxins are solid waste incinerators, cement kilns burning hazardous and non-hazardous waste, steel smelting and refining including iron ore sintering. Forest fires, which are increasing in frequency and intensity (an effect of human-induced climate change) are becoming a significant source.

29 United States Environmental Protection Agency *Summary of the Dioxin Reassessment Science* October 15, 2004 update.

30 World Health Organization, "Dioxins and their effects on human health" Fact sheet No. 225, May 2010.

31 *Final Report: Multi-Pollutant Emission Reduction Analysis Foundation for the Iron and Steel Sector*, prepared by Charles E. Napier Co Ltd for Environment Canada and the Canadian Council of Ministers of Environment, September 2002, pp. 137–9.

32 Ibid., p. 301.

FIGURE 9.1 Rolled steel produced by BHP Steel, Australia. Source: BHP Steel Co Archive (company no longer exists)

9 FUTURING: SUSTAINMENT BY DESIGN

Our previous chapters reviewed the variety of ways in which iron and steel-making have been implicated in ecologies of mind, matter and the social. We have endeavored to show how the industry's production processes, materials and products have significantly contributed to changing "natural" and made environments.

Iron and steel have become part of the very fabric of our taken-for-granted world to the extent of obstructing our ability to critically reflect upon what exactly these materials have created or destroyed. The environmental history of steel-making confirms the truism that nothing is created without something being destroyed. The future of the industry and our future increasingly rest upon improving our ability to see what our actions make and unmake, and thereafter make informed choices. It is in this context that practical questions and ethical decisions have to be made about the environmental costs and benefits of steel-making and steel products. In the last instance, unless steel has sustaining ability, it has no ability to justify its future production.

This final chapter will look to the future, but obviously not with a crystal ball.

The future cannot be viewed as a vacant space simply waiting to be filled by projected visions as utopians and naive futurists would have it. Nor is it appropriate to see it as replete with problems of unsustainability that in time science and technology will be able to resolve. Certainly, the future cannot be deemed as a coming age of renewed spiritual enlightenment. Rather, we need to think of the future as being in large part filled by the ongoing agency of things created in the past. The world-scape is littered with the evidence of human attainments, but equally with the enormous accumulated detritus of human errors. Unavoidably, human interventions in the "natural world" mean that each generation inherits things to manage, wastelands to remediate and the ever-growing challenge of striving to overcome the unsustainable.

Specifically, this chapter will address some of the factors that over-determine the agenda of advancing "sustainment" within the steel industry. It will then take a philosophical turn, directed towards informing the practical. To start with, two of the most pressing and linked cultural and material problems of the present are considered—climate change and unsustainable development.

Climate change

As discussed in the previous chapter, the IPCC has become increasingly confident that human activity is causing changes to global climate. More than 20 years ago, at the Rio Earth Summit of 1992, world governments started to talk about global warming and the need to reduce greenhouse gas emissions, agreeing that in order to avert climatic disaster in the twenty-first century, global warming should be kept below an average of two degrees Celsius. Governments kept on talking, and in 1997 at Kyoto, set modest greenhouse gas reduction targets of 5.2 percent for industrialized nations ("the Kyoto Protocol"). This target was, in fact, meant to be just a first move towards a reduction of 60 percent on current emission levels—the reduction that the world's leading climate scientists deem as needed. Yet the major emitters, USA and China have refused to commit to any reductions. Today, greenhouse gas emissions, rather than reducing compared to the 1990 baseline level, have increased by 50 percent.[1] 2012 saw the Arctic ice-cap melting at much faster than predicted rates, and climate scientists are now saying that due to failure to reduce emissions, we are possibly heading towards between 4 and 6 degrees global warming by the end of the century.[2]

In the face of the political paralysis of government, techno-pragmatists are promoting geo-engineering as the technological fix to the problem.[3] This prospect comes with two dangers: one is total unpredictability; the other is the possibility of action being taken by corporations independent of government control.

Clearly, this is an issue of major concern to the steel industry. It would be disingenuous not to acknowledge energy efficiency gains, process improvements to reduce pollutants, work on embodied energy assessment and life-cycle analysis—all of which have been advanced by the informed sections of the industry. Yet these modest reforms are doing little in the face of major problems—not least because the industry is still dominated by the goal of increasing overall output. That this equates to increased greenhouse gas emissions overall, seems to be ignored.

Policy responses to climate change encompass two courses of action: mitigation (to reduce the levels and impacts of greenhouse gas emissions) and adaptation (to deal with living in a still rapidly changing climate). What this division tends to obscure is that global warming and associated climate change are directly linked to the actions of producers and consumers: this through the volume of natural resources appropriated, how these are treated, what is made from them and the kinds of lifestyles that are inextricably bound up with specific patterns of resource and energy usage. Determining the appropriate extent of mitigation measures such as tree planting (to absorb carbon dioxide) is difficult because of the imprecision of the science of carbon sequestration. It is extremely hard to quantitatively correlate the relation between variations in ground vegetation, trees, soils and wetlands in the natural environment—what is clear is that different plant materials sequester carbon at different rates and over varied duration. Demarcating between "natural" and "human- induced" processes is also still problematic. In short, there is just not enough known about the behavior of carbon and the "carbon budget" of widely different ecologies.

The problem is accumulative. Because of the length of life of gases in the atmosphere (which can be more than 200 years) today's climate is a product of emissions from the nineteenth and twentieth centuries. So, even if the situation was to be stabilized soon via large and stringently applied emission reductions (extremely unlikely) climate change will still remain a factor for several hundred years. This is not least because it takes a very long time (several centuries) for mean ocean temperatures to adjust (ocean temperatures are the critical control factor in climate behavior).[4]

Against this backdrop, one has to act on trends rather than exact science, acknowledging the enormous risk from the ever-upward accumulation of greenhouse gases in the atmosphere. This suggests that there are enormous dangers in waiting for clear proof before taking decisive action.

Now to the other response to climate change—the need for adaptation.

Subject to global location, the human population is faced with very significant geo-climatic changes, which have already commenced. On the negative side—drought, more frequent extreme weather events (especially floods and landslides), higher UV levels, a gradual expansion of tropical climate regions (accompanied by a spread of tropical, vector-borne diseases), increased desertification, higher wind speeds (resulting not only in more wind damage but the loss of a great deal of top soil), rising temperatures (with corresponding increases in heat-related deaths, forest and grass fires) and changes in the life cycles of animals, insects and plants. Agricultural production and settlement will become impossible in some places. Because of these circumstances, it has been claimed that there are now 26 million environmental refugees, in contrast to 22 million from political conflict. On the positive side—the climate in some parts of the world will improve, summers and winters will be warmer and agriculture will flourish.

There will be major problems to confront, mostly in countries least able to deal with them. The choice here will be between financial and humanitarian aid or abandonment of whole populations (a trend, if one looks to Africa, already started). Second, and far more dangerous to global security, will be the proliferation of climate change refugees. Having large numbers of people on the move from inhospitable conditions and divested of their cultural grounding (climate is as much a cultural as a biophysical determinate) will pose threats to political stability arising from contestation over resources and the limited ability of many existing population centers to absorb large numbers of people. Again, this trend has already been recognized and the drift has started. Moreover, military strategists have designated it as a likely source of future wars—a language and a global mapping of potential environmental conflicts already exist (with water as the most politically "volatile" resource).

Where does the steel industry fit into this picture? Certainly, it faces major technical challenges to achieve significant emissions reductions (which we have addressed already in other chapters). Adaptation poses an even more enormous challenge, and goes directly to the relation between built forms, the environment and climate change. For there to be viable human future, in common with other environmentally high-impact industries, the steel industry does not have a future as it currently is.

According to global geography, buildings of all types, both old and new, as well as infrastructure, will have to deal with either more heat, rain, wind, snow, hail or dust. They will need more shade and external shelter; better management of their thermal mass; more insulation; greater ability to detain large volumes of storm water and discharge it at rates that do not exacerbate flash flooding; more durable landscaping; greater capability of withstanding wind speed and impact damage from hail and flying debris. Buildings that respond to this scenario will survive, but many won't. This situation will have profound structural design, construction and retrofitting implications. For example, standards will have to alter (again a fact just starting to be recognized by the International Standards Organization). Construction methods, facade and roof engineering will have to change, as will external plumbing. Various forms of storm water retention will add cost to building design and construction. In some areas, glazing will need more shade and wind protection devices.

As the twenty-first century progresses, climate adaptation and emissions reduction will become inter-linked drivers of building design. One can expect to see very different domestic, commercial, industrial and institutional architecture starting to appear in the coming decades. Clearly, this will have important implications for steel products in the built environment—some will be rendered obsolete, but many new ones will be needed.

Unsustainable development

We should consider briefly the overall model of "development," within which the varied patterns of usage of steel were established. A contemporary picture is already clear.

Levels of material "consumption" are growing as the global population increases and becomes ever more urbanized. The desire for "sophisticated" consumer goods and ecologically unsupportable lifestyles is accelerating, not least among "newly industrializing" nations. The flawed past of unsustainable late-modern/postmodern industrial nations still remains the model of the future for the "newly industrializing." What is lacking are alternative ideas, images and methods to enable poorer nations to leap from a pre-modern industrial (or dysfunctional) economy to a fully industrial one without replicating the error of the "advanced," quantity-based postmodern economies. As yet, the idea of sustainment is still not contributing to the creation of economies that can improve the quality of life and environment, and make a path to the future. As time will tell, the notion of growth-based development, undergirding "sustainable development," is incapable of solving the problems of uneven development and poverty.

These briefly outlined examples touch on a complexity that itself is a fraction of the complex problem of unsustainability.

Over two centuries of industrialization, plant and animal species have been depleted and biodiversity reduced. Industrially produced toxic chemicals, introduced into the environment by accident or intent, have transformed the biosphere over the last several hundred years.

With this situation, and all other examples of the unsustainable, we find a history of human-centered self-interests obscuring the ability to recognize that it is human values that are at the core of the problem of unsustainability. This condition cannot be overcome by an act of will of "the enlightened." Anthropocentrism is not something that is in addition to being human, but rather its essence. In this situation of non-transcendence, the idea of liberation is pointless. What is appropriate instead is accepting responsibility for what we are, and cultivating awareness of that for which responsibility has to be taken. Certainly the acceptance of this proposition should be regarded as an essential quality of leadership in almost every field of human endeavor.

Reactive measures such as quantifying pollutant discharges and monitoring biophysical impacts can tell us plenty about the symptoms of the unsustainable, but little of their fundamental causes. As discussed in the previous chapter, this kind of environmental reporting is the dominant approach of progressive organizations, including the more "enlightened" steel companies.

From the position outlined, we need to ask a range of questions of steel, critically engage "answers" and confront inscribed practices concerning: the *way* steel is made; *why* steel is used; *what* prefigures its use; *what* fixes the volume and form of products (including the technologies and social relations of production). The forcefully active and consciously back-grounded symbolic dimensions that determine how we perceive and act on the material require special attention. We have noted that in the past and present steel's iconic status (as product and industry) has been mobilized by individuals, organizations and nations. This is evident, commercially, in the images, design concepts, ideologies and technologies that underscore the market(ing) framing of its perception, use, valorization and disposal. Things are never exactly as they seem.

As matter or as an object of perception, steel cannot simply be reduced to the output of processes of production. What "it" is equally rests on a "community of meaning"—with its ability to link products, images, cultures, knowledge, environments and experience. Without such "commonweals" of shared sense and meaning, *our* world would simply fold into confusion and chaos.

The steel industry, in common with other industries, has more than an ecology and economy—it also has a social fabric. However, this fabric, and the community integral to it, is slipping away. To a large extent, technology has weakened the mutual dependence and trust formed by the dangers and hardship of labor; the skill and knowledge constituted and transferred in the workforce; the solidarity of the culture and the class created (and manifested) in times of industrial conflict or suffering. This—together with the continuity of traditions; sharing of pleasures; or a belief in a god, destiny or nation—frequently enabled attainments of mind and body to be reached, love to be unselfishly given and life to be sustained, all completely against the odds. While there is much of this past that was appropriate to displace, there is also a great deal to conserve that goes by unrecognized. Such conservation should not merely be for historical interest but for contemporary—and futural—needs.

The community formed in adversity, in opposition to the exploitation in the early part of the second industrial revolution was very different from the one sought to be created later by paternalistic employers attempting to artificially engineer community by the provision of housing and social structures to which all had to comply. Control of a workforce by authoritarian welfarism was epitomized by the rules and dependencies created by company housing estates—for instance, in the nineteenth century in England, by the Cadbury company (remembering that cocoa was a beverage deployed in alcohol abstinence campaigns) and in the twentieth century in the USA, by the railway rolling stock manufacturer, Pullman Palace Car Company of Chicago. One of the highest profile and most aggressive examples was Henry Ford's social project at Highland Park, Detroit. Here, in 1914, Ford set up a "sociology department" to administer a regime of surveillance and control of the moral behavior of workers after his introduction of the $5 day (which was then double the industry rate).

Learning and defuturing

While a comprehensive confrontation with the reality of anthropocentrism, the complexity of unsustainability and the elaboration of a new sovereignty is well beyond the scope of this chapter and the project of this book, it is crucial to understanding the directive force of steel's past, present and future and all that can be captured by the idea of "ecologies of steel." In other words, "the future of steel" has to be engaged within a planetary rather than in an industry context.

New technologies and reactive development

The search for, and realization of, technological advancement has existed from the very birth of iron and steel-making. Improvements in the manufactured material, production technology, applications, the lowering of manufacturing costs and, more recently, the reduction of negative environmental impacts have all been drivers of change. We have seen how technological advancements can become liabilities. For instance, once coke was a celebrated breakthrough, then coke-making became a problem and now perhaps its days are numbered.

All sorts of technological possibilities are proffered as the future of the steel industry. Some of the more immediate and developed ones were commented upon in Chapter 6. There are others on the more distant horizon such as using plasma energy to smelt iron at the mine site so that iron rather than ore is shipped out. Laser furnaces and nuclear fusion are other contemplated technologies. Certainly, one of the most basic lessons of the past is that no major technology should ever be introduced without a substantial accumulative impacts study, extremely comprehensive environmental management plan and a proper sign-off by an appropriately selected "precautionary principle" biased community of judgment. Again, the cost in time and money of such an exercise would be high, but the economic and ecological costs of not doing could be immeasurably higher.

More immediately, as we shall soon see with a few examples, the logic of production technology, materials recovery, reuse and recycling all invite exploration beyond the bounds of current practice. In this context, we can also talk about new materials. Let's start here.

The creation of new steels, super-steels, ultra-high-performance steels and the like is not only part of the research and development project of the industry, but, as we saw when looking at alchemy, represents a kind of thinking that is part of its culture and tradition. However, whole new areas of materials research are opened up by the imperative to advance sustainment by material "sustainments." At a basic level, the materials interfacing begs much greater exploration—not least the need to easily separate, clean and design-out the contamination of EAF steel by tramp materials (the quality of steel produced directly links to the quality of scrap). Changing the status and perceived values of materials is also part of the agenda of extended responsibility—a program of cultural reclassification could for instance have an enormously positive materials conservation outcome. In this respect "scrap" is an outdated and inappropriate naming which authors material neglect and misuse. Another ("jargon free") language is needed to communicate effectively to society at large the growing importance of material recovery and resource management—a term like "secondary resources" goes some way towards this.

We live in a contradictory age in which the immaterial forces of information and high-speed exchange of capital are overwhelming the significance of material forces in the popular imagination of most indus-trial/post-industrial nations. It is not that materials no longer count but, within the wider culture, they are becoming obscured from view and downgraded. Such an issue begs more recognition within the steel industry. Cultural strategies need to be explored and developed to go well beyond the more familiar concerns of "workplace cultural change" or "marketing steel with a positive cultural image."

Materials of invention, generic materials and the question of "sustainability"

As seen, steel has been, and still is, a material of invention. Any one of its forms at any given moment can be taken as a material from which to invent (the example of tool steel examined in Chapter 4 is a good illus-tration). As the market and environmental circumstances increasingly call for steel to be more "sustainably" produced, and have more "sustainable" qualities, and as the agenda of sustainment becomes more sophisticated, the pressure "to invent" and innovate will increase. Additionally, steel will have to find ways to respond to increasing competition from other materials, especially high-performance but un-recyclable composites and exotic metals like magnesium (which, to date, is still produced in very small quantities). Such materials are making inroads into steel's market share in the auto industry on the basis of reducing vehicle weight so as to reduce fuel consumption and thus greenhouse gas emissions.[5]

Alongside the unfolding contest of materials within particular industries, other changes are afoot. The very way we think about materials is starting to unravel. Rather than having fixed qualities, materials are now starting to be conceived of as infinitely variable, having the capacity to be made into whatever they are desired to be. Examples, some still experimental, include programming of genetic data to create a genetically engineered biodegradable material with a precisely designed lifespan; the mass production of change-in-process custom-designed polymers; the manufacture of high impact, ultra-high temperature super-hard ceramics; and the creation of new kinds of composites that uniquely combine the qualities of different types of materials.

These developments in the production of materials and the imperative of sustainment put on notice the notion of "core business," based on standard materials and the subordination of corporate direction exclusively to shareholder interests. Alongside "business as usual" is the possibility of an alternative future for steel emerging out of the industry's ability to survive, reinvent itself and flourish on the back of contributing to advancing the conditions of sustainment. One possibility here may well be the rise of the "inter-related products" (IPs).

A new model

The "inter-related products" IPs concept (which is not being claimed as unique) is based neither on the privileging of a single material nor is it simply centered on product diversification; rather its takes the production of multiple products, or co-products, as an organizational principle.

A steel company thus becomes a maker and marketer of a range of inter-related products that cluster around steel-making, well beyond the status of "byproduct-product." A product mix might be, for example, the existing range of standard steel products; new advanced alloy steels and products; reusable "standard structural steel components"; the sale of energy; BF slag cement and slag cement products (like bricks and pavers); liquid and solid waste management and engineered soils. Clearly, such a concept runs against the current wisdom of sticking to "core business" and is thus vulnerable to criticism from within the current economic paradigm. Certainly, the concept throws up many organizational issues and invites a great deal of investigation and creative innovation—driven by the many potential economic and environmental benefits.

Notwithstanding the challenge of numerous problems, and the need for a great deal of research, there is a clear possibility of reducing net steel production (the simplest method of reducing impacts), introducing supplementary products and making "extended (producer) responsibility" the basis of: new income generation; a new industry identity (from steel-maker to a materials maker, manager and "trading community"). Such a scenario would also change public perception of what the industry does economically, socially and materially.

In this model, steel could be manufactured by an automated continuous process with very low man-hours per ton ratio, while at the same time retaining, or even extending, a workforce via the "IPs sustainment value added products." This links to extending the structural position of the steel industry in supporting the development of the social ecology of the immediate and extended community, while also establishing a whole new set of inter-industry, local government and NGO relations. Likewise, such an approach could contribute to solving immediate environmental problems of waste management and greenhouse gas reductions—not just by making less steel, but by increasing the utilization of zero-rated materials (that is those waste or byproducts of steel-making, like slag, that have their emissions credited to steel) and by maximizing energy co-generation utilization.

Modest trends towards this kind of organization of production already exist. In Kalundburg, Denmark, for example, an oil refinery, a biotechnology company, a plasterboard manufacturer, a power-generator and a local authority have collectively created a micro-economy in which they trade energy and materials. For the steel industry, such cooperative structures could be mapped onto consolidations between corporations involved in competitor materials. The existing collaboration between steel corporations to create a "light-weighted" steel car body to compete with the increase of plastics, aluminium and magnesium in the auto industry is an example of this. The type of structural change signposted needs not to be simply subordinated to economic imperatives, but, as indicated, powered by the imperative of advancing sustainment, materially.

Of course, it is easy to find objections to such ideas. However, they beg to be tested against pragmatism based on: identifying opportunities; precisely defining problems; making critical choices; ability to deliver solutions; projecting likely consequences—all of this guided by located understandings of the biophysical, economic and cultural needs of sustainment. This adds up to a substantial change of direction that needs to be distinguished from the rhetoric of sustainability mobilized by those elements within the business community whose commitment to action does not go beyond "sustaining the unsustainable."

Beyond material and the immaterial

The relation between the material and immaterial is being redefined in the frame of sustainability.

One of the political and economic features of globalization impinging on the steel industry is the growing division between the rising star of clean *immaterial economic activity* and dirty *material industries*. The former are becoming a major characteristic of advanced economies, while there is a geographic drift of the latter to the newly industrializing economies. But this contrast is over-stated. What needs to be considered is what exists between and articulates the two poles of the material and immaterial.

Immaterialization (information, e-commerce, knowledge industries, software, financial services and so on) is not just a disengaged "other" of the material, but a supplement to it. Furthermore, the immaterial is a means of prefiguring the transformation of the material. For instance, material reuse as a systematic and general practice, while depending on embodied material factors, cannot occur unless it is immaterially prefigured (i.e. designed). Here, the designing idea (as concept, value, knowledge, change strategy, communication or logistics) is what drives the change that can reduce the overall volume of material produced and/or extend the life of material in use. This affirms the point made much earlier in the text, that an "ecology of mind" is relationally bonded to a material ecology. It follows that it is not possible to properly comprehend the "nature" of a material unless it is viewed from the perspective of the environment from which (its) meaning and use comes and goes. More than this, a material is not simply a collection of atoms, it is also a moment of embodied time, be it of variable duration. All matter is temporal and finite, even when its "life" is beyond our measure.

These comments point to the imperative to "extend the life of materials in use by immaterial means."

As implied, the meaning of a material can never be captured just by an exposition of its qualities and applications. Steel is no exception. A technocratic archaeology of its production cannot grasp its meaning adequately. No matter how much information is given, say on a material's forms, production, fixing methods, finishes or the products made from it, we learn very little. Would it be possible to comprehend a fine piece of architecture or an elegant object of engineering by just reciting the catalogue of materials and industrial process of its production? A recitation of the material composition of an object does not create a picture able to generate interest or concern. For this, we have to look elsewhere.

Immaterialization then can be recast as the prefiguration (by knowledge and design) of forms of the material rather than its disappearance. The entire software industry is predicated on this proposition—software drives hardware, the hardware exists simply to support it and what it thereafter does. The material is in effect "immaterialized" in order to be seen—we only see it via a screen of knowledge. One cannot "see" iron or steel as such without the knowledge to recognize it. Analogously, the idea of "sustainment" introduces a new knowledge that can transform how materials are seen and thought. It equally shifts how responsibilities for the material can be comprehended.

From a steel-maker's point of view, *information* can be used to determine material reuse. This could be extended to new models of material ownership. An immaterialization strategy could, for example, be of considerable importance in architecture and civil engineering. A move could be made to sell *material use* rather than end product. This would mean a steel-maker would retain material ownership, *but* lease the material's use, as well as provide design and technical support for efficient use and maintenance of the material in its functional use—all this to ensure recovery for reuse. Such "leasing-for-life-cycle" could fuse with and further extend the concept of "extended producer responsibility" (often understood as product "take back"); this could be implemented by legislation or by industry agreement within a broader notion of "extended *responsibility*."

In summary, the arguments and ideas in this chapter all strive to indicate that with thought and effort it would be possible to shift the center of gravity of steel companies away from an existing mind-set of "core business," conventional industry economics. Rather than lowering sights, what we have begun to conceptualize is an economy of much higher returns for investors, society and the industry itself; an economy that can substantially contribute to the sustainment of environments, communities and markets by remade organizational structures, production methods and immaterial strategies.

Design, redirective practice and time

The inclination towards "sustaining the unsustainable" is the dominant direction of existing product, engineering, architectural and information design practices. As they are now, they lack transformative ability. It follows that the nature of these practices demands to be transformed so as to acquire the agency of sustain-*ability*. The kinds of things design practices now need to know include:

- what of those designed things in their field of operation actually defuture and what products, services, knowledge, skills and practices have sustaining ability;

- how that which is designed and made acts to defuture;

- how to read the inscriptive power of objects, processes and services to sustain or defuture;

- how to recognize that products, services and systems are always relationally connected (and thus are process rather than a product);

- how to recover lost knowledge;

- how meaning is constituted, communicated and perceived; and how it can be reconfigured to transform material values and thus impacts.

This last point needs particular qualification. It is frequently assumed that sustainability automatically means material transformation; however, in many instances to change what something means transforms how it is viewed, valued and consequently how it is used. Impacts can increase or diminish according to shifts of meaning—that which was once waste becomes a resource, that which was neglected becomes cared for, that which was deemed to have a short life becomes long-lived. Changing meaning is thus a key means of sustainment.

Foregrounding the importance of changing meanings and creating "redirective practices" reconnects with the proposition that the unsustainable is anthropocentrically located in "us"—thus the need to engage values, desires and self-centered actions. However, this requires one additional and substantial qualification. The making of new meanings and the creation of redirective practices takes time—it takes time to make time, remembering that making the sustainable is a making of time. In the last instance, the bottom line is not economics but sustainment—the precondition for every modality of (biological and economic) exchange.

Design innovation and new ultra-standards

The UltraLight Steel Auto Body—Advanced Vehicle Concept (ULSAB-AVC) discussed in Chapter 6, is a design strategy that illustrates change in process, design-led innovation and a certain pragmatic engagement with sustainment. Although such developments are directionally and incrementally positive, they are still very much within the remit of "sustaining the unsustainable" and thus vulnerable to substantial criticism. While light-weighted car bodies can reduce fuel consumption, they do little to engage the major problem of the sheer number of cars on the planet. Additionally, there is the escalation of gridlock in most of the world's major (and many minor) cities—which itself generates many economic, social and environmental problems. Thus, the rapid growth of car ownership erases any gains in emission reduction from light-weighting even if it were to become the norm. So while the action partly mitigates the problem, it does nothing to solve it.

A less problematic example of light-weighting is the way high-performance steels (HPS) have allowed the design and construction of bridges requiring less material, less maintenance, having a longer life and lower cost.[6] Over 14 such bridges have been built in the USA, which itself has generated a demand for more. There are still problems to confront with these structures, especially the relation between stressing, weight and wind. Equally, the function of a bridge within an entire transport infrastructure and the nature of that infrastructure itself, all invite interrogation.

Affirmatively, however, the use of high-performance materials to reduce materials output is a positive example of how economic viability can be retained or extended, while production, and its associated environmental impacts, can be reduced. In many respects such thinking is not new; at the same time, approaching it with knowledge of ecological interconnectedness adds the possibility of a whole new design and development process. There are also innovations—recalling earlier remarks on "materials interface" and "inter-related product and materials"— which suggest quite new uses and relations between materials. Current research into the use of high-performance fiber-reinforced polymer decks for steel bridges would be one example of this.

Another design strategy that has made a mark in both product design and architecture, especially in terms of steelwork fabrication, is "design-for-disassembly" (DFD). The idea is to enable "direct" or "adaptive" reuse by deploying either traditional fixing methods like bolting or new kinds of fasteners, so as to facilitate rapid disassembly. DFD aims to deliver by design a multi-life and/or multi-function building or a building with reusable system components that extend the life of material in use (which is a superior, lower impact outcome than materials recycling). The economic shift implied by DFD is from income based on the sale of materials to income from smart design services.

Such approaches change evaluative norms—both performatively and aesthetically. They change how capital investment is able to be viewed (capital cost, payback period and rate of return over time, etc.) and in so doing increase the scope and importance of design tasks.

Other developments are possible. Neo-standardization and ultra-standards that reflect a changing climate are design strategies with potential for sustainment.

Neo-standardization is a possible trans-industry design project in which a comprehensive design regime is developed, based on "standardized reusable structural components" like beams, girders, compression members (struts, columns), tension members (bars, tube, angle), connection plates, structural framing (purlins, girts). These would all be system-designed so as to lift prefabrication and DFD to another level of design. It could combine information management (e.g. information on "building-life management" and component identification) with a highly systematized regime of standardized components and a

DFD ethic to create the possibility of very flexible multi-life structures. This revisits and reinvents one of the founding principles of the second industrial age (an age which paradigmatically established the economic "logic" of volume output by mass production) which was the design and manufacture of interchangeable parts. The contemporary challenge is to enable the "the designing of difference" with standard components. This requires highly tuned design skills, innovative design software and the rigorous development of a comprehensive range of components with industry agreements on new global standards for size and tolerance. The "imperative of sustainment" needs to be the common value that makes the negotiation and collaboration possible.

Neo-standardization opens the way to ultra-standards—standards for the design and manufacture of products accredited for multiple life applications. Clearly, there are technical research implications in setting metallurgical standards for the production of steel, components and structures able to deliver a required performance over extensive periods in changing conditions.

A much greater use of design concepts, services and management together with new services such as the recovery of reusable components (rather than just "scrap"), their testing, certification, storage and resale—all of this can be viewed as potentially viable activity of a more sustainable steel industry in which "extended responsibility" is a fundamental marketing base. Recycling is of course not rejected—the reverse: it has to be improved.

These kinds of product/service mixes would, of course, be linked to different pricing structures. A cost differential between production for reuse and for custom manufacture would need to be established and generally applied—products being rated on their sustainment value, on the basis of a multi-life-cycle assessment cycling, rather than just on scrap recovery value.

Public perception and trust

As has been pointed out, a great deal of the rhetoric of sustainability folds into sustaining the unsustainable. While many actions undertaken in the name of sustainability are vulnerable to this criticism, there is a need to distinguish between actions taken in good faith, backed by a desire to reduce damage to the environment, and those that are mere "green-washing"—the cynical use of environmental gestures and rhetoric predicated on placating or deflecting criticism on poor environmental performance. Green-washing is little more than a play of representational appearances. Like many industries, the steel industry has had its share of "green-washing." At the same time, the industry has often been confronted, especially in "developed" economies, with having to deal with the large environmental signature of steel works—steel has consequently been fingered as one of industrial society's biggest polluters. Such characterizations have not emerged in the abstract, but usually against the backdrop of a steel company whose environmental, social and economic presence looms large over a town or city. For the more responsible companies, green-washing "public relations" has given way to relations with the public that recognizes environmental action for the common good. Public consultation, more adequate reporting and community partnerships have become essential measures to maintain the goodwill of both the public and the market. Investments have been made to lay foundations of trust upon which to build. Such action, which is still significantly undeveloped, is not only a reflection of change but essential for it.

While the public can perceive the steel industry as a problem, it is also seen by the populations of steel towns and cities as part of the fabric of their culture, community and local economy (in many instances economic dependence being a significant factor). Without doubt, affirmative perceptions and community attachments are elemental to those social ecologies that need to be sustained. There is a major point to be made here, one that any good manager would know well—winning the trust and support of the local community is an increasingly important corporate asset.

Industry leadership towards sustainment comes at a large cost; change is hard (especially in an epoch when the rhetoric of change has become exhausted, when change seems to be constant, and its direction is unclear, abstract or far distant). So said, change towards sustainment is essential—it is the primary "essentialism of living now." This means stopping what is known to be unambiguously harmful, it means knowing what should be done and finding out how to do it, it means making a fundamental effort to create new options, it means vision, hard thinking and courage, it means educating shareholders, management, workers and clients—all via a managed process of transition.

The norm of modernity has been for technology to lead change. However, as our whole project evidences in a variety of ways, sustainment, while able to be technically assisted, cannot arrive simply by technological means. It requires cultural transformations to establish, extend and inscribe those meanings, values, actions and attitudes so as to make sustainment culturally elemental, and thus something that permeates our education, occupation and recreation. All of these remarks are made to just set up one observation: to change, the steel industry needs help, and potentially the most powerful and important source of that help is from its own communities. Leadership in this context is about asking not imposing. It's about honestly *saying what has to change and how change can come by constituting and supporting a community of change.*

In "developed economies" the move to sustainment does mean a fall in the standard of living (judged according to current norms); it does mean the establishment of redistributive justice as a basis for world trade and it does mean significant alterations to lifestyle. This is the "high cost." The trade-off, the gain to be created, the argument to be waged, is that the shift from a quantity economy to a quality (and far more sustainable) economy is able to bring about healthier populations, as well as happier and more meaningful ways of working and living.

A last word on sustainment

Throughout this book, three aims have dominated: first, turning the eyes and minds of the steel industry outwards towards the historical, futural and relational complexity of what we have called the ecologies of steel; second, redirecting designers', engineers' and architects' understanding of the steel industry and its sustaining potential; and finally, showing all interested constituencies that steel is always situated in relation to other materials and forces that articulate the worlds we inherit, occupy and constitute.

Our enterprise has travelled with a type of thinking and a formative "ecology of mind" directed towards a sustainable future that is based on reading and learning from circumstances of the present and selectively drawing upon an historically interrogated past—be it of innovators, technologies or the values and practices of other cultures. It is worth considering, for instance, that past innovators were not specialists but "relationists" with a highly developed sense of past and present contexts. To take two whose significance has already been established: Sir Henry Bessemer and René-Antoine Ferchault de Réaumur.

Bessemer, a professional inventor, was born in rural England in 1813, the son of French refugees. While he was a man of demonstrably great creative, technical and inventive powers, he was not averse to appropriating the ideas of others.[7] Besides his acclaimed attainment in metallurgy—the Bessemer converter—he "invented" many other things: movable stamps for dating deeds and other government documents; an improved typesetting machine; "gold" powder made from brass for use in paints; sugar cane-crushing machinery; a solar furnace; an astronomical telescope; and machines for polishing diamonds. His encounter with steel-making in 1854 was the consequence of trying to find an appropriate metal for a rifled barrel for a gun he had invented for the Crimean War.[8] Réaumur, in contrast, was an eminent French scientist and foremost entomologist of the early eighteenth century who conducted research in varied fields. He devised the thermometric scale bearing his name, published multiple volumes of what were to become seminal works of entomological history and researched human biology, with particular focus on digestion.[9] Here is the backdrop against which we can view his improved techniques for making iron and steel—including his development of a cupola furnace. This knowledge, registered in his seminal work on metallurgy, *L'Art de convertir le fer forgé en acier*, was published in 1722.

We should remind ourselves of the ancient Chinese attainments in steel-making technology that both of these men rediscovered and advanced.[10] Besides this common denominator, we also note with considerable interest that Réaumur was a scholar of Chinese science and technology, in particular the chemical composition of Chinese porcelain—an area not exactly distant from the concerns of other pyrotechnic arts, structures and refractory materials.

Our focus on the ecologies of steel has been via the imperative of *sustainment*, this term having been adopted to create differentitation from the limited (industry) view of "sustainability," which by and large centers on the environmental impacts of production and the life-cycle impact materials.[11] To reiterate, sustainment is learning to think and act otherwise. Some of its fundamental aspects, summarizing from what was said at the beginning of this chapter, are:

- that nothing can be sustained without an identification of the unsustainable;

- that all that is unsustainable stems, in the first instance, from "our" values, desires and actions (human anthropocentrism);

- that how, what and why we unsustainably produce our material world needs to be much better understood—specifically, "the way and the why" steel is used and what prefigures its use (including the symbolic dimension);
- that the sustainment of individuation rests on the commonality of community.

In summary and conclusion

In the face of the absolute imperative of sustainment and the defuturing character of the steel industry, change is not a matter of choice but necessity—be it that such change requires to be seen within the context of a wider transformation of industry at large. In terms of it environmental impact it has to become much smaller—clearly this can only happen if the demand for steel is significantly reduced. For this to happen, industrial production and rapid urbanization has not only to slow but directionally change. Less, but better, goods need to be made and smaller structures built. Likewise, as indicated, the steel industry has to become more diversified (thus dramatically reducing waste) and a more important provider of services and solutions. In other words it has to earn income in new and less materially intensive ways. Clearly all of this means an enormous amount of technological and organizational innovation, together with a major change in the culture of the industry. While such action poses massive problems (not least because they are indivisible from much large and wider changes in the world, and because the industry is not a homogenized whole) it is nonetheless appropriate to demand and expect timely action and leadership. To say this is, in part, to recognize that in the end there is no choice. The industrial leviathan will destroy the ecologies of human dependence if it does not change. As the slogan goes: "adapt or die." Obviously an industry has to have a transformative process. It has to build a platform of change within the organization that allows for continuity while the form and process of change is being created. What this means is a continual shift to the new mode of the industry and a constant diminishment of the old.

Notes

1   Economic growth in China, other parts of Asia, South America and
    Africa are the reason for the increase. While Kyoto Protocol signatories
    reduced their emissions collectively by 16 percent, this was due not to
    virtuous actions, but to the collapse of industries in Eastern Europe
    and the recent global economic crisis. In 1990 the US accounted for
    two-thirds of global emissions, now it contributes less than 50 percent.
    Since 2000 carbon dioxide emissions in China have nearly tripled. But
    this has to be seen in the context of the migration of heavy industry from
    developed to developing countries which make products that get shipped
    to wealthy nations. Quirin Schiermeier, "The Kyoto Protocol: Hot air"
    Nature, November 28, 2012. www.nature.com/news/the-kyoto-protocol-
    hot-air-1.11882 (accessed December 12, 2012).

2   Dr. Jim Yong Kim, President of World Bank wrote in his Foreword to a
    recent report: "It is my hope that this report shocks us into action. Even
    for those of us already committed to fighting climate change, I hope it
    causes us to work with much more urgency. This report spells out what
    the world would be like if it warmed by 4 degrees Celsius, which is what
    scientists are nearly unanimously predicting by the end of the century,
    without serious policy changes. The 4°C scenarios are devastating:
    the inundation of coastal cities; increasing risks for food production
    potentially leading to higher malnutrition rates; many dry regions
    becoming dryer, wet regions wetter; unprecedented heat waves in many
    regions, especially in the tropics; substantially exacerbated water scarcity
    in many regions; increased frequency of high-intensity tropical cyclones;
    and irreversible loss of biodiversity, including coral reef systems. And most
    importantly, a 4°C world is so different from the current one that it comes
    with high uncertainty and new risks that threaten our ability to anticipate
    and plan for future adaptation needs." *Turn Down the Heat: Why a 4
    degree Celsius warmer world must be avoided: A Report for the World
    Bank* by the Potsdam Institute for Climate Impact Research and Climate
    Analytics, Washington, DC: International Bank for Reconstruction and
    Development/The World Bank, November 2012.

3   See Tony Fry "Living in Darkness" in *Becoming Human by Design* London:
    Berg, 2012, pp. 179–97.

4   As the IPCC's Fifth Assessement Report states: "A large fraction of
    anthropogenic climate change resulting from CO2 emissions is irreversible
    on a multi-century to millennial time scale, except in the case of a large
    net removal of CO2 from the atmosphere over a sustained period. Surface
    temperatures will remain approximately constant at elevated levels for
    many centuries after a complete cessation of net anthropogenic CO2
    emissions. Due to the long time scales of heat transfer from the ocean
    surface to depth, ocean warming will continue for centuries. Depending

on the scenario, about 15 to 40 percent of emitted CO_2 will remain in the atmosphere longer than 1,000 years." Working Group I, *Contribution to the IPCC Fifth Assessment Report Climate Change 2013: The Physical Science Basis Summary for Policymakers*, September 27, 2013, p. 20.

5   The introduction of these materials has prompted the steel industry towards innovation, and within the auto industry, light-weighting projects. Perhaps the best known example is the ULSAB-AVC (UltraLight Steel Auto Body—Advanced Vehicle Concept) project from 1994 to 1998, in which over 30 steel-makers in Asia, Europe, South Africa, North and South America participated. It aimed to create a new "advanced steel automotive body architecture" for vehicles that would be safe, strong, affordable and significantly lighter. Light-weighting of materials enables improved "power to weight ratio" to deliver a light car with a small engine equal in performance to a heavier car with a much larger engine (resulting in much lower fuel consumption and thus reduced emissions). The engineering of the project was led by Porsche Engineering Services based in the USA and the outcome was to achieve a body that weighed 36 percent less than the mid-size sedans benchmarked. This kind of design activity falls very much into the "reactive" frame—not least to the inroads being made by the aluminium industry into what seemed for a long time to be uncontested steel territory. One of the project's key points was stated thus: "The new paradigm is that steel has evolved as a lightweight material. We are challenging the auto industry to think of steel when they think lightweight, and throw out the paradigms of steel past." American Iron & Steel Institute website www.autosteel.org/ulsab_avc/index.htm

6   See Brian Fortner "Forging Ahead" *Civil Engineering* April 1999, pp. 60–1.

7   See L. T. C Rolt *Victorian Engineering* Harmondsworth: Penguin Books, 1970, pp. 182–3. Rolt gives an account of Bessemer's failure to acknowledge and reward Robert Mushet, who had the metallurgical knowledge he lacked for solving the problem of excess oxygen in the steel made in his early converter, this being vital knowledge to make marketable product.

8   Ibid., p. 182.

9   See entries in *Britannica* CD, Version 99 © 1994–9. Encylopædia Britannica, Inc.

10   As indicated in Chapter 1, Needham described the Chinese "hundred refinings" method of steel-making as "theoretically ancestral to Bessemer conversion" and observed "direct migration of Chinese workman skilled in this work immediately preceded the group of inventions associated with the name Bessemer." Joseph Needham *The Development of Iron and Steel Technology in China: Second Biennial Dickinson Memorial Lecture* London: Newcomen Society, 1956, p. 47.

11   Here the concept of sustainment is subordinate to a larger understanding (the Sustainment), which is a major intellectual project equal to the one

that underpinned the making of the modern world (the Enlightenment). Essential Sustainment is the project of what the modern world has brought into being. See Tony Fry *Becoming Human by Design* London: Berg, 2012, pp. 143–62.

INDEX

ecology 3, 24, 49, 124, 141, 154, 160, 166, 179, 182, 183, 185, 188, 192, 195, 197, 201n., 219, 233, 238
of meaning 9
of mind 11, 24–7, 29, 37, 28, 42n. 25, 69, 71, 75, 78, 79, 98, 105, 127, 133, 144, 238, 245
social 109, 112, 145, 237
of steel 24, 69
economics/economy 1, 4, 5, 8, 13, 17, 38, 40n. 1, 51, 55, 59, 60, 62, 66n. 18, 86–8, 97, 98, 104–9, 119, 120, 126, 134, 136–8, 143, 148, 149, 153, 159, 160–4, 167, 170n. 2, 171n. 10, 172n. 12, 177, 180, 185–7, 193, 194, 203n., 204n., 207, 209, 211, 213–14, 220, 223, 232–4, 237, 238, 240–5, 248
Eggert, Gerald, G. 204n. 43
Egypt 21, 71
Elba 18
electric arc furnaces *see* furnaces
Elizabeth I 55
emissions control 11, 54, 59, 62, 63, 67, 88, 128, 129n., 153, 166, 167, 176, 185, 192, 193, 196–8, 207, 209–11, 215, 217–23, 228, 230, 236, 237, 248n. 1, 249n. 5
Emrich, Walter 67
Enfield Armoury 123
engineering 3, 12, 19, 71, 86, 88, 93, 100, 104, 113, 117, 119, 122–4, 128, 130n, 145, 147, 149, 212, 213, 228, 231, 239, 240, 249n. 5
genetic 4
engines 32, 36, 45, 64, 100, 104, 114, 120, 147, 156n., 184, 187, 196
blowing 32
English 23
Enlightenment 34, 69, 71, 75, 78, 107, 227, 250n. 11
environment(al) 2–4, 6–9, 11–14, 39, 50–2, 55–7, 59, 60, 62, 64, 66n. 18, 67n. 38, 74, 80, 87–9, 97, 98, 104, 105, 108, 113, 128, 129n. 7, 134, 135, 137, 144, 149, 153,
154, 159–61, 163, 165–7, 169, 176–89, 191–5, 197–9, 201–4n., 207–21, 223, 225n. 31, 227, 229–35, 237, 240–9
refugees 230
ethics/ethical 4, 7, 165, 227
Ethiopia 23
Eurocentrism 17, 19, 43, 49, 106, 141
Europe/European 5, 6, 11, 16, 19, 20, 22, 25, 26, 28, 34, 35, 37, 42n. 28, 51–60, 71, 72, 91, 98, 99, 100, 101, 105, 106, 119, 121, 137, 140, 143, 144, 147, 148, 150, 164, 170n., 183, 186, 189, 201, 208, 215, 248n. 1, 249n. 5
exhibitions 151, 152, 165

fabrica 136
ferric oxide 195
Ferrill, Arthur 155
first industrial age 145
Fisher, Douglas, A. 42
flexible manufacturing systems 127
Ford, Henry 127, 233
Forest of Dean 52, 114
Fortner, Brian 93n. 43, 249n. 6
fossil fuels 49, 62, 168, 198, 216, 217
foundries 6, 65, 118, 138, 142, 188, 211
France 52, 71, 99, 123, 143, 149, 170n. 8, 171n. 12
Fry, Tony 10, 14n. 4, 41n. 1, 66n. 28, 131n. 47, 172n. 15, 174, 248n. 3
furnaces 6, 21, 22, 27, 32, 34–6, 42n. 18, 44n. 43, 45n. 48, 52, 64, 65n. 9, 66n. 21, 72, 77, 80, 92n. 39, 99, 100, 103, 110, 111, 116, 120, 121, 122, 126, 170n. 8, 172nn. 11, 12, 175, 176, 185, 188–91, 199, 212, 213
basic oxygen (BOS) 67n. 40, 176, 199, 219
blast 20, 21, 27, 28, 33, 36, 43n. 31, 45nn. 42, 48–50, 54, 58–60, 62, 64, 67n. 38, 83, 84, 92n. 38, 99, 103, 104, 107, 109, 121, 129n. 6, 163, 167, 168,

sustainable/sustainability 6, 8–10, 40n. 1, 165, 179, 207, 234, 238, 240, 243, 246
sustaining the unsustainable 134, 166, 168, 238, 240, 241, 243
sustainment 7, 8, 10, 12, 13, 18, 40, 40n1, 168, 169, 228, 232, 235–47
Sweden/Swedish 36, 52, 54, 58, 60, 65n. 12, 67n. 20, 82, 99, 120, 142, 181, 215, 216
  College of Mines 66
  Royal Institute of Technology 60
symbol/symbolism 2, 4, 7, 9, 18, 22, 23, 69, 70, 76, 89n. 1, 91n. 20, 98, 107, 153, 159, 162–5, 232, 247
systems 2, 5, 12, 34, 36, 38, 39, 50, 52, 56, 58, 65n. 9, 71, 78, 112, 120, 123, 127, 128, 143, 144, 148, 149, 153, 191, 193, 195, 199, 208–10, 213, 217, 218, 238, 240, 242, 248n. 2
Szechuan 20

tantalum 81, 91n. 34
Tao/Taoism 50, 71, 73
Tartars 38
Taylor, Frederick Winslow 115, 125–7, 131nn. 38, 39, 40, 42, 46
technology/technological 4–6, 12, 13, 20, 22, 23, 26, 31, 32, 36, 37, 39, 40, 43n. 38, 45n. 46, 46n. 60, 58, 59, 60, 63, 64, 67n. 36, 80, 81–6, 97–104, 106–9, 112–14, 119, 121, 123, 127–9n. 7, 136–8, 141, 143,146, 148, 151, 154, 155n. 15, 156n. 20, 160–9, 170nn. 3, 8, 171nn. 11, 12, 194, 199, 210, 212, 214, 216, 219, 222, 227, 228, 232–5, 237, 244–8, 249n. 10
Telford, Thomas 43n. 33, 150
tempering 28
Temple, Robert 46n. 57
temples 39, 151
Thales 75

Thames 54
Their, Dave 91n. 30
thermochemical 11, 35, 57, 64, 72, 80, 83
Thomas and Gilchrist 35
Tonkin, Cameron 14n. 2
towers 150, 197, 219
Trevithick, Richard 146
Tribune Building (1873) 157n. 34
Trossero, M. 67n. 32
Turkey 37, 43n. 24, 165

Ukraine 223
unsustainable/unsustainability 4–6, 8–10, 12–14n. 4, 40, 49, 56, 57, 89, 127, 128, 134, 160, 161, 166, 168, 169, 177, 227, 228, 231, 232, 234, 238, 240, 241, 243, 246
urban /urbanization 40, 148, 149, 153, 161, 163, 167, 187, 188, 208, 211, 213, 218, 231, 247
Ure, Andrew 21, 42n. 15, 92n. 37, 103, 191
US
  Army Ordinance Department 85
  Bureau of Labor 203n. 43
  Clean Air Act 215, 218, 219, 220
  Clean Water Act 219
  EPA 204n. 44, 218
  Marines 93
  Navy Ordinance Bureau 85
  Research Office 80
  Toxic Release Inventory 219
utopia/utopias 13, 127, 128, 227

Voest-Alpire AIRFINE 222
volatile organic compounds 196

Walker, R. D. 171n. 8
war/warfare 8, 12, 18, 19, 23, 80, 86, 92, 106, 120, 131n. 36, 133, 135, 136–8, 140–5, 148–50, 152, 154, 155nn. 2, 8, 9, 13, 17, 164, 165, 170nn. 2, 8, 171n. 12, 186, 199, 230, 246
water 20, 21, 25, 29, 33, 37, 43n. 41, 49, 50, 52, 56, 58, 62, 62, 66n.